VLSI PLACEMENT AND GLOBAL ROUTING USING SIMULATED ANNEALING

THE KLUWER INTERNATIONAL SERIES IN ENGINEERING AND COMPUTER SCIENCE

VLSI, COMPUTER ARCHITECTURE AND DIGITAL SIGNAL PROCESSING

Consulting Editor
Jonathan Allen

Other books in the series:

Logic Minimization Algorithms for VLSI Synthesis, R. K. Brayton, G. D. Hachtel, C. T. McMullen, and A. L. Sangiovanni-Vincentelli
ISBN 0-89838-164-9.
Adaptive Filters: Structures, Algorithms, and Applications, M.L. Honig and D. G. Messerschmitt.
ISBN 0-89838-163-0.
Computer-Aided Design and VLSI Device Development, K. M. Cham, S. -Y. Oh, D. Chin and J. L. Moll.
ISBN 0-89838-204-1.
Introduction to VLSI Silicon Devices:Physics, Technology and Characterization, B. El-Kareh and R. J. Bombard.
ISBN 0-89838-210-6.
Latchup in CMOS Technology: The Problem and its Cure, R. R. Troutman.
ISBN 0-89838-215-7.
Digital CMOS Circuit Design, M. Annaratone.
ISBN 0-89838-224-6.
The Bounding Approach to VLSI Circuit Stimulation, C. A. Zukowski.
ISBN 0-89838-176-2.
Multi-Level Simulation for VLSI Design, D. D. Hill, D. R. Coelho
ISBN 0-89838-184-3.
Relaxation Techniques for the Simulation of VLSI Circuits, J. White and A. Sangiovanni-Vincentelli
ISBN 0-89838-186-X.
VLSI CAD Tools and Applications, W. Fichtner and M. Morf, Editors
ISBN 0-89838-193-2
A VLSI Architecture for Concurrent Data Structures, W. J. Dally
ISBN 0-89838-235-1.
Yield Simulation for Integrated Circuits, D. M. H. Walker
ISBN 0-89838-244-0.
VLSI Specification, Verification and Synthesis, G. Birtwistle and P. A. Subrahmanyam
ISBN 0-89838-246-7.
Fundamentals of Computer-Aided Circuit Simulation, W. J. McCalla
ISBN 0-89838-248-3.
Serial Data Computation, S. G. Smith and P. B. Denyer
ISBN 0-89838-253 X.
Phonologic Parsing in Speech Recognition, K. W. Church
ISBN 0-89838-250-5.
Simulated Annealing for VLSI Design, D. F. Wong, H. W. Leong, C. L. Liu
ISBN 0-89838-256-4.
Polycrystalline Silicon for Integrated Circuit Applications, T. Kamins
ISBN 0-89838-259-9.
Fet Modeling for Circuit Simulation, D. Divekar
ISBN 0-89838-264-5.

VLSI PLACEMENT AND GLOBAL ROUTING USING SIMULATED ANNEALING

by

Carl Sechen
Yale University

Kluwer Academic Publishers
Boston/Dordrecht/London

Distributors for North America:
Kluwer Academic Publishers
101 Philip Drive
Assinippi Park
Norwell, Massachusetts 02061, USA

Distributors for the UK and Ireland:
Kluwer Academic Publishers
Falcon House, Queen Square
Lancaster LA1 1RN, UNITED KINGDOM

Distributors for all other countries:
Kluwer Academic Publishers Group
Distribution Centre
Post Office Box 322
3300 AH Dordrecht, THE NETHERLANDS

Library of Congress Cataloging-in-Publication Data

Sechen, Carl, 1956–
VLSI placement and global routing using simulated annealing/by Carl Sechen.
p. cm.—(The Kluwer international series in engineering and computer science ; 54)
Bibliography: p.
Includes index.
ISBN 0-89838-281-5
1. Integrated circuits—Very large scale integration—Design and construction. I. Title. II. Series; Kluwer international series in engineering and computer science ; SECS 54
TK7874.S38 1988 88-12889
621.395—dc19 CIP

Printed in the United States of America

Table of Contents

Preface

From my B.E.E degree at the University of Minnesota and right through my S.M. degree at M.I.T., I had specialized in solid state devices and microelectronics. I made the decision to switch to computer-aided design (CAD) in 1981, only a year or so prior to the introduction of the simulated annealing algorithm by Scott Kirkpatrick, Dan Gelatt, and Mario Vecchi of the IBM Thomas J. Watson Research Center. Because Prof. Alberto Sangiovanni-Vincentelli, my UC Berkeley advisor, had been a consultant at IBM, I received a copy of the original IBM internal report on simulated annealing approximately the day of its release. Given my background in statistical mechanics and solid state physics, I was immediately impressed by this new combinatorial optimization technique. As Prof. Sangiovanni-Vincentelli had suggested I work in the areas of placement and routing, it was in these realms that I sought to explore this new algorithm.

My first implementation of simulated annealing was for an island-style gate array placement problem. This work is presented in the Appendix of this book. I was quite struck by the effect of a nonzero temperature on what otherwise appears to be a random interchange algorithm.

My initial objectives were to develop high performance, industrial strength algorithms for standard cell and macro/custom cell placement (and floorplanning). In order to compare the quality of placements, one requires both global and detailed routers. The lack of good global routers for these layout styles inspired me to develop both standard cell and macro/custom cell global routers.

Despite the title, only about one-half of this book directly concerns simulated annealing. The remaining topics, however, arose strictly as a result of seeking to apply simulated annealing to automatic placement. The standard cell placement algorithm (Chapter 3) and the macro/custom cell placement algorithm (Chapter 4) are based on simulated annealing. The standard cell global router described in Chapter 3 does not use simulated annealing.

Chapters 4 through 8 are all concerned with the macro/custom cell layout style. The overall placement algorithm is presented in Chapter 4. In macro/custom cell placement, it is necessary to estimate the interconnect space needed between the cells. A new method for *dynamically* estimating the interconnect area around each cell is introduced in Chapter 6. That is, the estimated interconnect area associated with a cell is a function of its position.

This interconnect-area estimator, intended for use in conjunction with a simulated annealing based placement algorithm, requires an accurate estimate of the *final* total interconnect length at the *beginning* of the simulated annealing algorithm. This requirement inspired the development of models for accurately predicting the average interconnection length for both random and optimized placements, described in Chapter 5.

In macro/custom cell placement, after the execution of the simulated annealing algorithm, there is a need to refine the placement based on more accurate estimations of the required interconnect area. This is accomplished by performing channel definition and subsequently global routing. In Chapter 7 I introduce a channel definition algorithm which is, to the best of my knowledge, the first such algorithm suitable for the placement-refinement problem. This algorithm is also useful for generating channels intended for detailed routing in the macro/custom cell layout style.

Chapter 8 presents a new graph-based global routing algorithm. Typically the graph would be generated by means of a channel definition algorithm such as that described in Chapter 7. The key approach in the new algorithm is to first generate many alternative routes for each net and subsequently to choose a particular route based on the minimization of channel congestion and chip area. This strategy largely avoids any dependence on the routing order of the nets.

I am very grateful to Prof. Alberto Sangiovanni-Vincentelli for making it possible for me to pursue the research presented in this book. His standards of excellence will always be a model for me.

I am much obliged to many UC Berkeley colleagues, particularly Peter Moore, Rick Spickelmier, Tom Quarles, Ken Keller, Res Saleh, Rick Rudell, Fabio Romeo, Andrea Casotto, Jeff Burns, and Prof. Richard Newton.

Carl Harris of Kluwer deserves special mention for suggesting this book and subsequently for encouraging its completion.

I would like to thank my students in the Yale CAD group, namely Dahe Chen, Tom Khamkoon, Kai-Win Lee, Mindy Lee, and William Swartz for their help in debugging and proofreading the entire manuscript. Several Yale students also deserve thanks for proofreading portions of the manuscript, in particular, Kamal Khuri-Makdisi, Lee-Chung Lu, and Joel Yang.

I would like to extend my deepest appreciation to Annie Yih, whose contributions ranged from technical to compositional, including proofreading. Your personal advice was a vital source of inspiration.

Aside from a total of four paste-ups, this book was typeset exclusively on a Macintosh SE by Apple Computer. Excellent software products by Microsoft (Word), Allan Bonadio Associates (Expressionist), Cooke Publications (MathWriter), Silicon Beach (SuperPaint), Cricket Software (Cricket Draw), and Apple Computer (MacDraw) made it all possible. The thrill of never having to use *troff* (or other batch word processors) again cannot easily be expressed in words.

The writing of this book would not have been possible without the generous support provided by Digital Equipment Corp., Texas Instruments Inc., Lawrence Livermore National Laboratory, Intel Corporation, and the Semiconductor Research Corporation.

List of Figures

List of Tables

Chapter 1

Introduction

1.1 Placement and Global Routing of Integrated Circuits

Presently an integrated circuit (IC) can have a feature size on the order of one micron and the IC can measure up to 10 mm on a side. Furthermore, current processing techniques require between 8 and 14 mask levels. Hence, the layout of an IC can have more than 10^9 mask edges. Human design of an IC of this complexity would clearly require a prohibitive amount of time. Further, the probability of making mistakes would be very high. One method of reducing this enormous complexity is to use the technique of *divide-and-conquer*. In this strategy, a large problem is recursively sub-divided into smaller problems until their size becomes manageable. For example, in a large random logic digital integrated circuit, a three-input NAND gate may be used hundreds of times. Using a divide-and-conquer strategy, the designer can carefully perform the layout of a pair of adjacent three-input NANDs and then store the result in a *cell library*. Each time one or two three-input NANDs are required to be laid out, the designer can simply copy the layout from the cell library. In this fashion, a manual layout step need be applied only once for each unique cell type. Full characterization, including verification and simulation, is similarly performed only once for each cell in the library. Furthermore, when designing a subsequent integrated circuit, the cell library may be re-used.

While the creation of the cell library does substantially reduce the complexity of the IC design process, the total number of *cell instances* in a single IC can be as high as 50,000 today. If it were of no consequence as to where a cell is *placed* on an integrated circuit, and if it were of minimal concern as to how the cells were electrically interconnected, or *routed*, then this level of complexity could be handled by human designers. In fact, for different cell placements, the area of an IC can vary by a factor of three or more. Since the IC production cost is at least exponentially related to the chip (IC) area, the cell placement is a major concern.

For some types of digital integrated circuits, the amount of area available for interconnect is strictly limited. If there is not enough interconnect area available for a given cell placement, then a new placement must be generated. This time-consuming process must be iterated until all of the interconnect can be placed. The *wireability*, a measure of how difficult it is to complete the interconnect for an IC, is a very strong function of the cell placement.

Circuit performance is yet another major concern in the design of digital integrated circuits. If the delay through a particular signal path is excessive, the circuit may not perform in accordance with the logic specification. For integrated circuits, signal delay results in part from the lengths of the interconnections on the chip. Hence, if the maximum signal delay is to be minimized, then the length of the interconnections along the *critical paths*, that is, the signal paths whose delays limit the overall performance of the chip, must either be minimized or kept below a certain length. The conclusion is then that the placement of the cells and the routing of the interconnections for those cells are of primary importance in the layout of digital integrated circuits.

Given N_C cells on a chip, the number of possible placements is of the order $O(N_C!)$. For $N_C \approx 50{,}000$, it is an impossible task for a human designer to find the best possible placement. The possibility of applying the fastest possible computer to this task is also not promising. The placement problem, in which it is desired to find the placement of minimum total interconnect length, belongs to the class of problems known as NP-complete.[1] In fact, most placement and global routing problems belong to the class of NP-complete problems. For problems in this class, there is no known exact algorithm whose worst-case time complexity is bounded by a polynomial in the size of the input. Consequently, heuristic algorithms are used to find solutions to these problems. These algorithms explore a discrete space of admissible configurations S in a deterministic fashion. Starting from an initial configuration $i_0 \in S$ a sequence of configurations are explored until a satisfactory one is found. The algorithm is specified by the rules according to which configurations are generated and accepted. Often the search terminates in a *local minimum*, at a state $\hat{j} \in S$ such that $c(j) \geq c(\hat{j})$ for all $j \in \hat{S}(\hat{j})$, where $c()$ is the cost function defined on the configuration space and where $\hat{S}(\hat{j})$ represents the states of S reachable from $\hat{j}$. The state $\hat{j}$ is a local minimum if there exists a state $\bar{j} \in S$ such that $c(\bar{j}) < c(\hat{j})$. In many cases, the quality of the local minimum states, or configurations, are quite inferior in comparison to the *global minimum* state. Heuristic algorithms often end at these local minimum states because their behavior is essentially *greedy*, that is, only those states which maximally reduce the cost are accepted.

Consequently, prior to 1982, the heuristics developed for automatic placement generally performed very poorly in comparison to human placement. Automatic placement was not widely used in industry, even for circuits of 1000 cells and beyond, which were

---

[1] M. Garey and D. Johnson, "Computers and Intractability: A Guide to the Theory of NP-Completeness," (1979).

considered large circuits at the time. Another problem contributing to the lack of utilization of automatic placement algorithms was their inability to capture many of the basic requirements and flexibility demanded by industrial layout problems.

Global routing, sometimes termed loose routing, is the preliminary step of the complete routing process. The goal of global routing is to assign each net to particular regions on the IC which are reserved for routing. Not only is it desired to complete the assignment for every net, but it is also of importance to minimize the total interconnection length of all of the nets. The latter objective is often used in attempts to minimize the IC area and to maximize the IC electrical performance.

Typically global routers operate on a channel graph, or routing region graph, where each edge represents a routing channel and where each edge has a cost associated with its length and with the remaining channel capacity. The nodes in the graph represent intersections of channels. For many layout styles it is therefore possible to pose the global routing problem in terms of a graph theoretic problem. Because of the fact that automatic global routers can often operate independent of many of the IC design rules associated with the interconnect layers, they have been successfully used for several layout styles. On the other hand, adequate global routing algorithms have not been developed for all layout styles.

A common problem for global routing in many layout styles is to find the shortest path which interconnects the pins belonging to a single net. Several algorithms are available to find the minimum interconnection length for two-pin nets, for example, Dijkstra's algorithm[1] is useful for global routing on a graph and the Lee-Moore algorithm for grid-based global routing.[2] In this latter global routing style each coordinate, or grid, on each interconnect layer in the IC either is, or is not, available for routing a specified net in which each pin is located at a particular grid location. The minimum length route on a graph for an n-pin net belongs to the class of NP-complete problems. This implies that it is inherently a difficult problem to solve. For the case of global routing one net at a time, there have been some heuristics developed which appear to give reasonably good solutions. However, the difficulty with many of the methods developed to date is that the quality of the solution depends in a significant way on the order with which the nets are routed.

In contrast, detailed routing is at the other extreme. Because of the inherent strong dependence on the IC interconnect design rules, until quite recently automatic detailed routing has been performed only on the least dense layout styles. These styles usually permit the detailed routing to follow some type of *grid*, be it real or virtual. The grid spacing is usually the worst case design rule, such as the contact-to-contact pitch. This grid spacing is far too pessimistic for full custom, high volume chips such as microprocessors. Automatic routing has only recently been used extensively for the custom case, due to the development of some very flexible routers.

The remainder of this chapter is organized as follows: Section 1.2 presents the cell-based placement and global routing problem. Previous approaches to the cell-based place-

---

[1] E. Dijkstra, "A Note on Two Problems in Connection with Graphs," (1959).

[2] C. Y. Lee, "An Algorithm for Path Connections and its Applications," (1961).

ment and global routing problem are presented in Section 1.3. Finally, in Section 1.4, the new approach to the cell-based placement and global routing problem is described.

1.2 The Cell-Based Placement and Global Routing Problem

Currently, the most common cell-based IC layout styles are *gate array*, *standard cell*, and *macro/custom cell*. In general, the layout configuration, or *image*, is different for each of these styles. The placement problem for each of these images has in common the *core region* into which the cells are to be placed. This is shown in Figure 1.1. The core region is bordered by pads. Typically the ordering of the pad placement is specified by the designer of the integrated circuit, although this is not universally true.

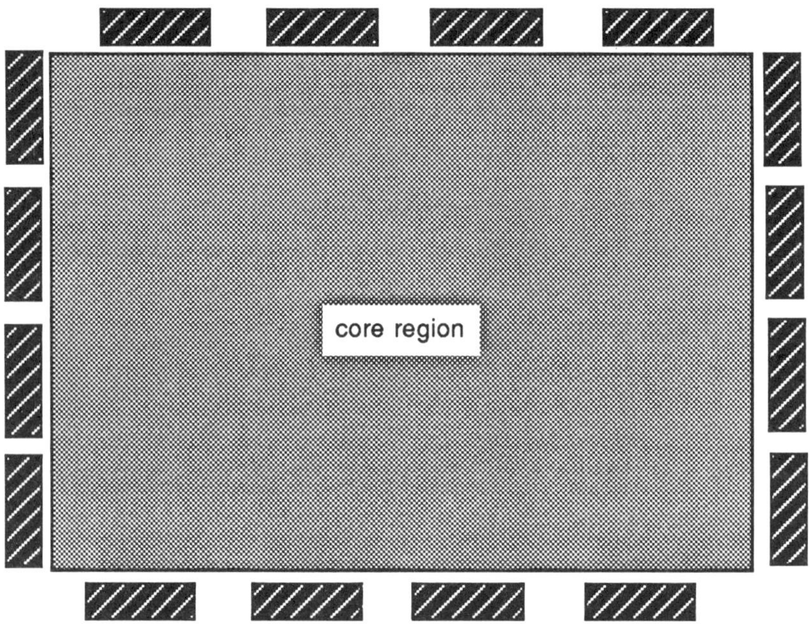

Figure 1.1 Generic image for cell-based layout.

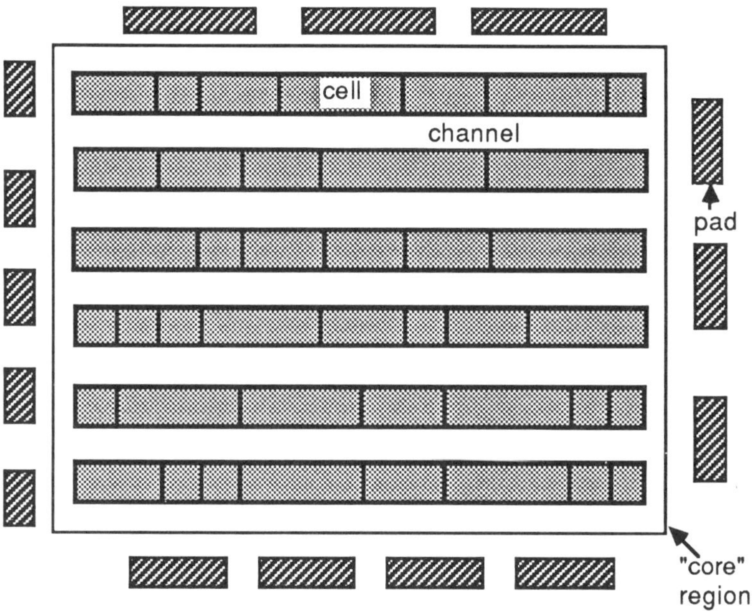

Figure 1.2 Simplified gate array and standard cell layout images.

1.2.1 The gate array placement and global routing problem

Gate array and standard cell images are typically both row-based layout styles. An example of a simple gate array or standard cell image is shown in Figure 1.2. A number of different images are used for gate array designs.[1] A predefined pattern of transistors is fabricated on a gate array chip. All of the fabrication steps are completed with the exception of the metal layers and the metal-to-metal contacts. Thus, the personalization of the chip is effectively restricted to the metallization steps. Current technology is such that two, and sometimes three, layers of metal are used for interconnection. The density of an IC (the number of transistors per unit chip area) is directly related to the number of metal layers. On the other hand, the degree of difficulty in terms of fabrication is strongly related to the number of metal layers.

[1] W. DeCamp, G. Sporzynski, and H. Burbank, "Gate Array and Standard Cell Approach," (1986).

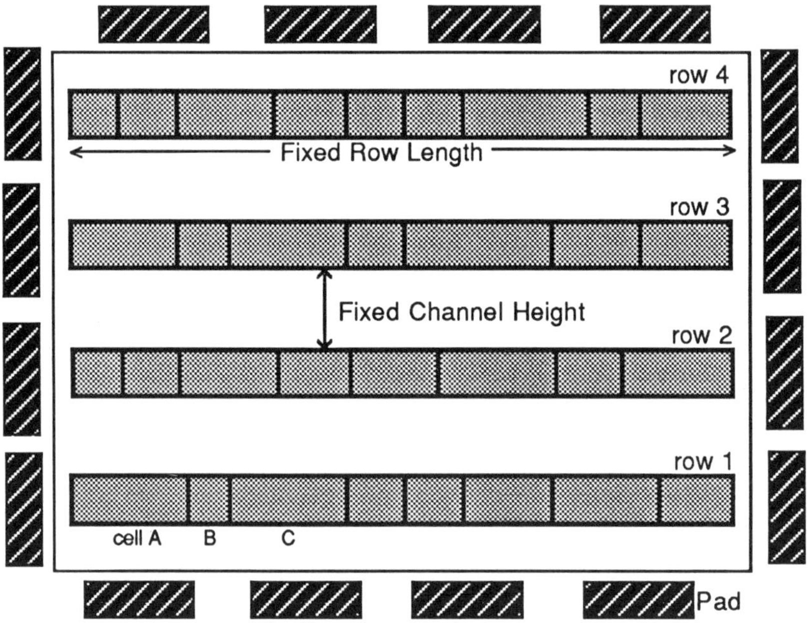

Figure 1.3 A row-based gate array layout image.

Since every chip contains the same pattern of transistors, regardless of its function, the chips can be manufactured in volume. The metal personality is added only on demand for a particular function. The unique part of the processing is only a portion of the total manufacturing time. Hence the time from design to hardware is significantly reduced. Further, debugging of the fabrication process need only be done once for a large variety of chip functions. Because of this latter point the technology used for gate arrays is closer to the leading edge.

An example of a row-based gate array layout image is shown in Figure 1.3. Because the fabrication has been completed through the silicon levels, the horizontal channel heights are fixed, as are the row lengths. Figure 1.4 shows a topographical layout of a typical cell.

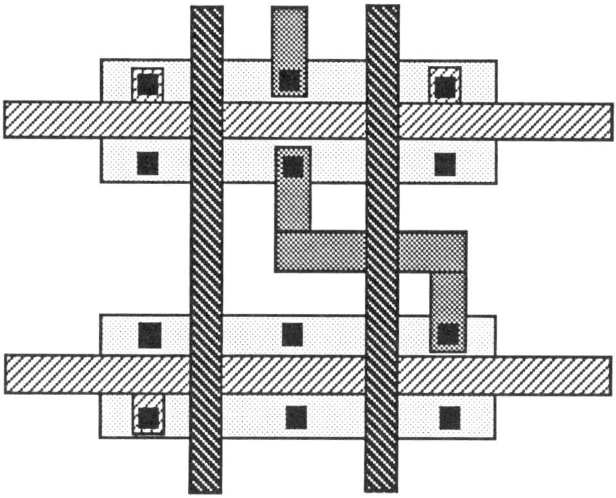

Figure 1.4 A topographical layout of a typical cell.

This cell is a 2-input CMOS NAND gate. All input and output connections of this gate are brought to the cell boundary next to the channels. Cells performing different functions are often made the same height. The VDD and GND lines are typically at the same vertical positions. In this fashion, it is not necessary to explicitly route VDD and GND; they are interconnected by abutment. This is illustrated in Figure 1.5. For the purposes of automatic layout, the VDD and GND signals can be ignored until the vertical side channels are routed.

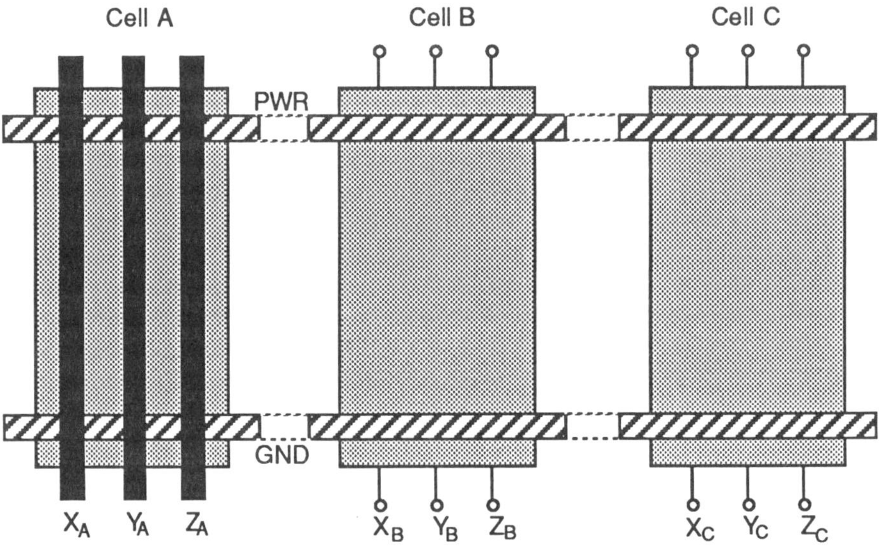

Figure 1.5 Row-based layout style: PWR and GND lines are interconnected by abutment. The interconnect layer used for PWR and GND signals is different from the interconnect layers used for the inputs and outputs (X, Y, and Z).

The placement and routing model for gate array, as well as other row-based images, is shown in Figure 1.6. Essentially, the cells are indicated as rectangles of common heights and various widths, and are arranged in rows. The terminal or pin positions are indicated as points within the cell boundary. Most often, these points are located on the boundary of the cell. The pins are given a label which is the name of the signal or net to which the pin belongs. A set of pins having the same label must be electrically interconnected.

Gate arrays shorten and simplify the design process. However, this is done at the expense of silicon efficiency. The designer of the gate array image must anticipate all possible uses of the gate array. Hence, the transistors must be made larger than if they were designed for a particular application. Furthermore, in a typical design, it is not possible to use all of the transistors. This is one source of wasted silicon area. Another is that the channel heights must be selected in consideration of the anticipated demand for horizontal interconnections across a vertical cut line. The selection of the channel heights must also be made in anticipation of the highest gate utilization. That is, those circuits which require most of the prefabricated transistors. Of course, the required channel heights are also a strong function of the placement and routing tools available. In terms of selecting the channel heights, it is an advantage if the same tools will be used for all designs, and this is often the case. In any event, largely due to the variable gate utilization, it is usual that almost all channels have greater heights than necessary, resulting in additional wasting of silicon area.

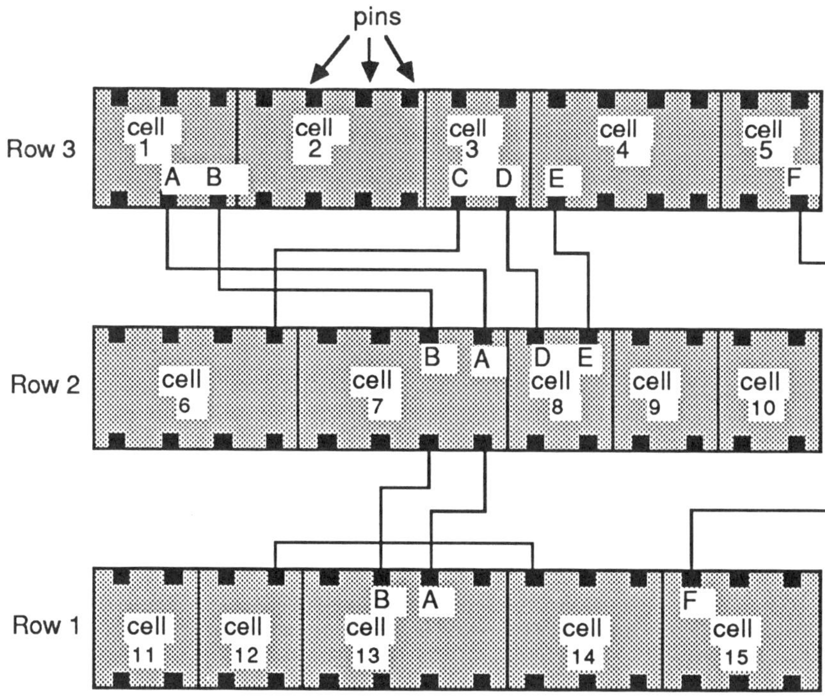

Figure 1.6 Placement and routing model for row-based images.

The larger transistors result in greater power usage than a customized design. Also, the switching speeds of gate array transistors are slower than those in a customized approach. For a given logic function, circuit performance is inversely related to the chip size. The inability to tailor transistors for a given application also impacts performance. This is offset somewhat by the fact that the technology for gate arrays is usually closer to the leading edge in comparison to a customized design style. In gate arrays, it is often true that improved technology permits smaller transistors and hence better chip densities. In summary, gate arrays offer a significant advantage in design time and design cost. On the other hand, the performance and power suffer, with some of the cost benefits eroded by the wasted silicon area.

The main goal of automatic placement and routing is to be able to interconnect every signal net within the area defined by the wiring channels. If this is not possible, then human intervention is required to attempt to complete the routing. This manual process does not always succeed in completing the interconnections, in which case the cell placement must be altered. A secondary goal, having satisfied the first, is to maximize the chip performance.

Another gate array image, termed an *island-style gate array,* is shown in Figure 1.7. An island-style gate array is such that an array of cells covers the chip with pads on the periphery. Each fundamental unit in the gate array is referred to as a *cell.* Hence, a 50×50

gate array is said to have 2500 cells. Island-style gate arrays have more complex cells than the simple row-based gate arrays. These cells are area efficient for integrated circuits with a large number of complex circuit functions. Since the cells have many more pins, or terminals, than a row-based cell, some of the pins are placed on the sides of the cells. The initial connections to these pins are made in the vertical channels.

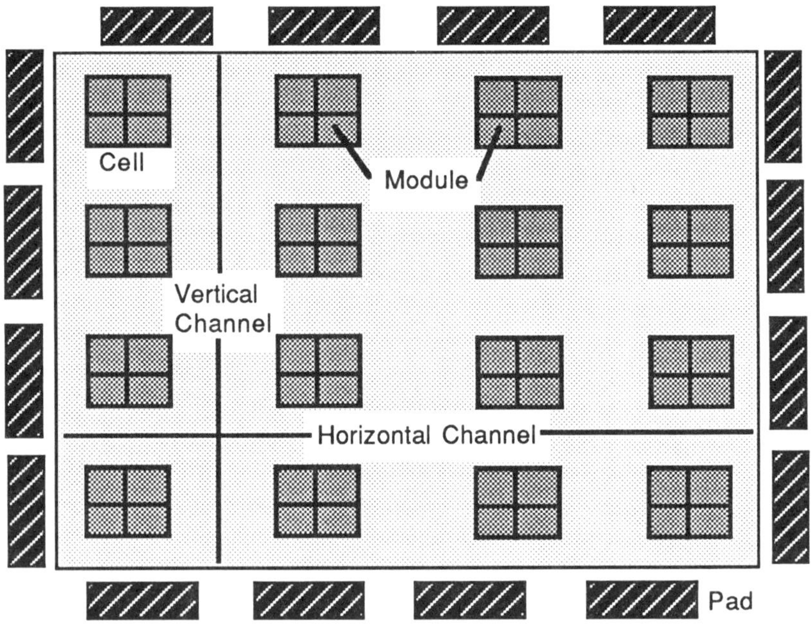

Figure 1.7 Island-style gate array image.

In an island-style gate array, functionally independent units are often contained within a cell. For example, some island-style gate arrays have functional units which are full-cell, half-cell, or quarter-cell sized. The term *module* is used to refer to a fundamental unit specified in the net list. In this example, a module may be the size of: (1) a full cell, (2) a half cell, or (3) a quarter cell. The placement problem for this image is to find a unique location within the array of cells for each module such that all of the interconnections will *fit* in the vertical and horizontal channels.

1.2.2 The standard cell placement and global routing problem

The generalized standard cell layout style is characterized by horizontal rows of standard cells with pads placed around the periphery of the chip. Furthermore, macro blocks may be present on the chip. An example of such a standard cell layout is shown in Figure 1.8. In fact, the standard cell image is very similar to the row-based gate array image. The primary difference between the standard cell and the gate array approaches is that standard cells are personalized on all mask levels. This permits the addition of macro blocks on the same chip.

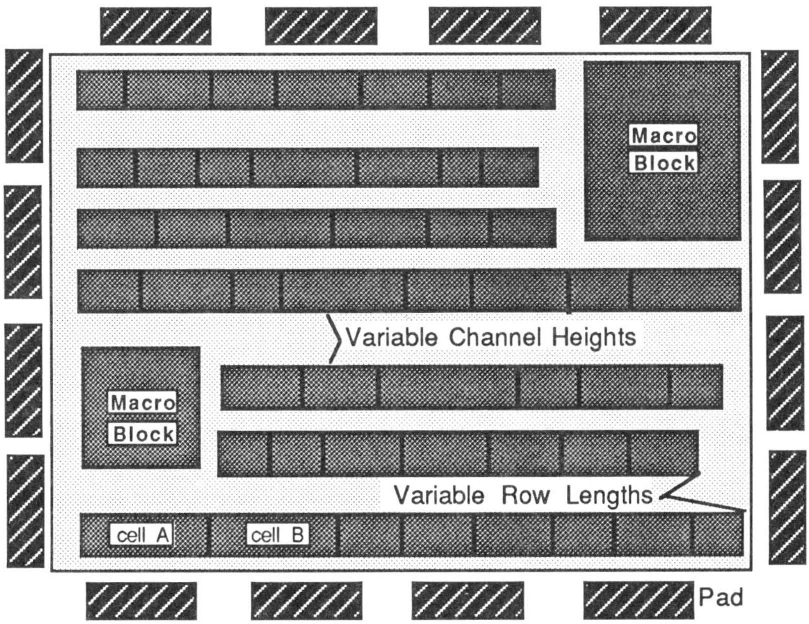

Figure 1.8 Generalized standard cell layout image.

The standard cell approach results in greater silicon efficiency in comparison to gate array since only those transistors which are needed in a circuit are fabricated. In addition, the transistors which are used can be tailored to the specific usage. The ability of incorporating macro blocks on the same chip enables more complex circuits to be implemented in standard cell designs.

The placement and global routing problem is somewhat different for standard cell images. Since all fabrication steps remain to be performed, it is always possible to achieve a 100-percent routing completion rate in a standard cell layout. This is because the channel heights are made exactly as large as necessary to accommodate the interconnect. Consequently, the primary goals for automatic placement and global routing tools are to maximize circuit performance and to minimize chip area.

1.2.3 The macro/custom cell placement and global routing problem

Macro/custom cell is the most dense, the most flexible and the highest performing design approach. In general, the cells are permitted to have any rectilinear shape. Furthermore, the cells may have fixed geometry including pin locations (macro cells) or the cells may have an estimated area with a specified aspect ratio range, and with pins that need to be placed (custom cells). Custom cells may have aspect ratios in a continuous or discrete range. Furthermore, several alternative instances for a custom cell may be specified. The selection of aspect ratio and/or instance is guided by the maximization of chip performance and by the minimization of the chip area. Individual circuits may consist of a combination of macro and custom cells. Figure 1.9 illustrates an example of a macro cell image. Pin positions are indicated on the cell edges. Both horizontal and vertical channels are used to route the interconnections, in the manner indicated.

Since all fabrication steps remain to be performed, it is always possible to achieve a 100-percent routing completion rate in a macro/custom cell layout. The channel separations are made exactly as large as necessary to accommodate the interconnect. The primary goals for automatic placement and global routing tools are to maximize circuit performance and to minimize chip area.

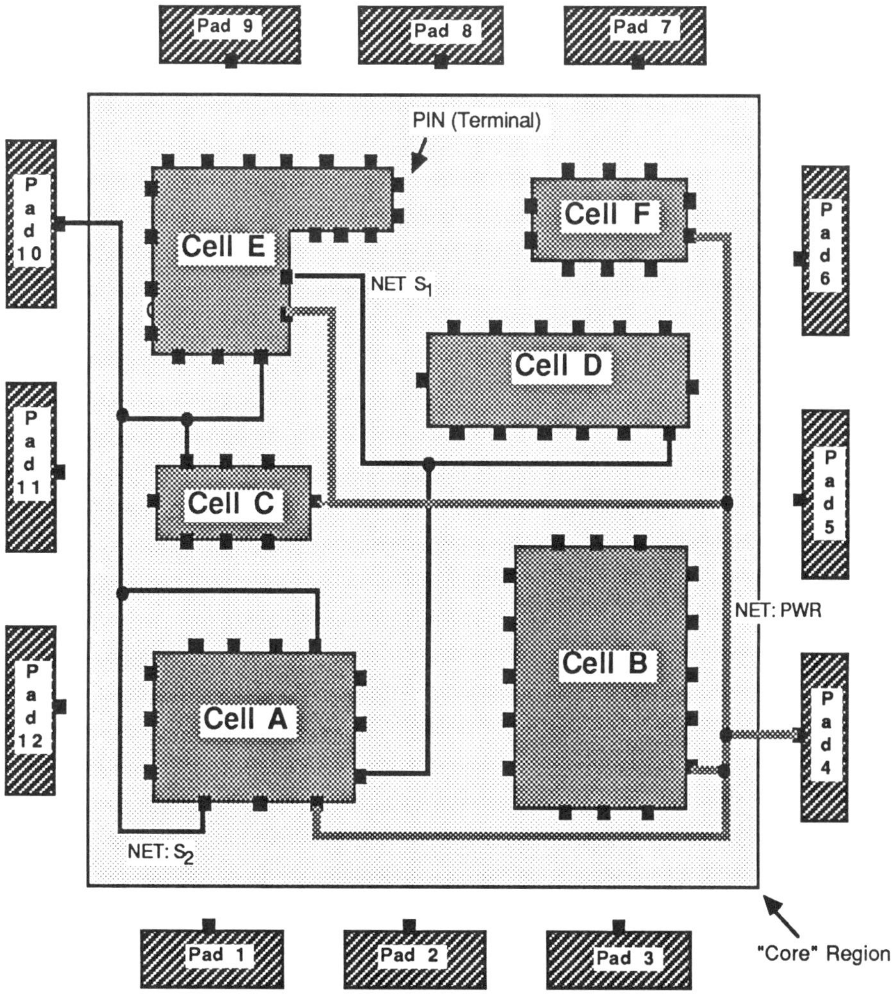

Figure 1.9 Macro/custom cell layout image.

1.3 Previous Approaches to Placement and Global Routing

1.3.1 Previous placement methods

A summary of the performance of early automatic placement techniques was published in 1976.[1] In this time frame, the placement models were quite simple. All cells were assumed to have the same size and the image was assumed to be a rectangular array of

[1] M. Hanan, P. Wolff, and B. Agule, "Some Experimental Results on Placement Techniques," (1976).

cell sites, or slots. The various placement algorithms which were compared have been described in detail.[1] The algorithms which were studied included: (1) constructive initial placement, (2) pairwise interchange, (3) neighborhood interchange, (4) Steinberg's Algorithm,[2] (5) force-directed relaxation, (6) force-directed interchange, and (7) force-directed pairwise relaxation.

The constructive initial placement algorithm selects cells, one at a time, based on an evaluation function which measures the degree of net connectivity to cells already placed. The selected cell is then placed on an available position on the image such that the total net length to the previously placed cells is minimized.

The pairwise interchange method begins with an initial cell placement. Then, two cells are selected and are trial interchanged. If the interchange does not result in an increase of the total net length, then it is accepted. If the interchange increases the total net length, the cells are restored to their initial positions. If there are n cells, then $n(n-1)/2$ interchanges are attempted in a *cycle*. If the total net length is not decreased in a cycle, the procedure ends. In the neighborhood interchange method, cells are only interchanged with nearby cells. A cycle is completed when each cell has been interchanged with each of its neighbors. Again, if the total net length is not decreased in a cycle, the procedure ends.

Steinberg's algorithm operates on independent sets of cells, that is, cells which have no nets in common. An iteration consists of: (1) selecting an independent set of cells, (2) removing the set of cells from the image, and (3) computing the cost of placing each of the cells in each available slot. Since these costs are independent of the location of the other cells in the independent set, an optimum assignment of cells to slots can be found by solving the resultant linear assignment problem.

In the force-directed algorithms, a *force vector* $\vec{F}_k$ is computed on each cell k, where

$$\vec{F}_k = \sum_i c_{ki}\vec{d}_{ki} \tag{1.1}$$

and where c_{ki} is the edge weight connecting cells k and i, and $\vec{d}_{ki}$ is the vector distance from k to i. This force vector is used to compute the target location for cell k, that is, the location where the sum of the forces on the cell is zero. When k is placed in the slot which minimizes the force, the cell which it displaces is *relaxed* next. In the force-directed interchange method, the force on a cell k is first computed. Cell k is then trial interchanged with its three neighbors in the direction of the force vector.

Another algorithm based on force vectors is the force-directed pairwise relaxation method. A cell a is chosen and its zero-force slot is identified. A cell b is identified which lies within a distance ε of the target slot for a and whose own zero-force slot is within a distance ε of the original location of a. Cells a and b are then trial interchanged. If the total net length is not increased, the interchange is accepted. Based on experimentation with small circuits, Hanan, Wolff and Agule concluded that the force-directed pairwise relax-

[1] M. Hanan and J. Kurtzberg, "Placement Techniques," (1972).

[2] L. Steinberg, "The Backboard Wiring Problem: A Placement Algorithm," (1961).

ation method performed the best among the above techniques.[1] This method is appealing since the probability of acceptance for each proposed interchange is very high.

The generalized force-directed relaxation method was described more recently by Goto and Matsuda.[2] In their method, the interconnect length for a net is approximated as one half of the perimeter of the minimum rectangle, or *bounding box*, enclosing the pins comprising the net. When a cell a is selected for relaxation, this cell is temporarily deleted from all of the nets which connect to it. The bounding box is then recomputed for each of these nets. The goal is to replace the cell such that the total increase in the spans of the bounding boxes is minimized. Since each direction can be treated independently, it is sufficient to consider only the x direction. The left and right edges of each bounding box are sorted. Suppose there are $2N$ such edges, where N is the number of nets connected to the cell to be relaxed. Then the x span between edges N and $(N+1)$ constitutes the *median* of the cell for the x direction. Taking into account both directions, the median is such that when the selected cell a is replaced into the nets and the bounding boxes recomputed, then the total increase in the spans of the bounding boxes is a minimum.

The second step in the process, represented by $\lambda = 2$ where λ is the step number, is to select ε cells in, or around, the median of cell a. Each of these cells, in turn, is temporarily placed in the original position of cell a. If at least one of the interchanges results in a reduction in the total routing length, then that interchange which results in the greatest reduction is accepted. If none of the interchanges results in a reduction, then cell a is placed at the location of cell b among the ε cells such that the increase in the routing length was a minimum. The median of cell b is then computed. The step characterized by $\lambda = 3$ continues the process with the selection of ε cells in, or around, the median of cell b. Each of these cells, in turn, is temporarily placed in the original position of cell a. If at least one of the interchanges results in a reduction in the total routing length, then that interchange which results in the greatest reduction is accepted. If none of the interchanges results in a reduction, then cell b is placed at the location of the cell c among the ε cells such that the increase in the routing length was a minimum. The third cell c, which was displaced by cell b being placed in its median, is then relaxed. The process then continues with $\lambda = 4$. In practice, λ is limited to a maximum value thereby implying that the original cell a is not moved if a reduction in total routing length could not be found in λ steps. If no cell exists which can be moved with λ steps, the placement is said to be λ-optimum. The results yielded by this method seemed to be comparable to those given by force-directed pairwise relaxation, for nominal values of ε and λ.[3]

---

[1] M. Hanan, P. Wolff, and B. Agule, "Some Experimental Results on Placement Techniques," (1976).

[2] S. Goto and T. Matsuda, "Partitioning, Assignment, and Placement," (1986).

[3] S. Goto and T. Matsuda, "Partitioning, Assignment, and Placement," (1986).

In 1970, Kernighan and Lin published a partitioning algorithm.[1] The Kernighan-Lin algorithm became the basis of several published placement methods.[2] [3] [4] [5] The objective in partitioning is to divide a circuit into two parts such that the number of nets which connect to cells in each of the two parts of the circuit is minimized. The number of such connections is referred to as the *cut* value of the partition. The original Kernighan and Lin algorithm assumed that every net connects exactly two cells. Furthermore, it was assumed that every cell has the same size. The method assumed an initial arbitrary partition of the $2N$ cells into two sets, A and B. When two cells are interchanged, the *increase* in the cut value is represented by the gain g.

The algorithm begins with the selection of two cells, $a_1 \in A$ and $b_1 \in B$, such that the interchange of these two cells results in the minimum gain g_1, (that is, the maximum reduction) in the cut value. The two cells, a_1 and b_1, are then interchanged and they are forced to remain in their new sets for the remainder of the current pass of the algorithm.

The second step is to select two cells, $a_2 \in A$ and $b_2 \in B$, such that the interchange of these two cells results in the minimum gain g_2 in the cut value. The two cells, a_2 and b_2, are then interchanged and locked in their new sets for the remainder of the current pass. These steps continue until all cells are locked. At this point, the cells originally in A are now in set B and those originally in set B are now in A. This marks the end of the current pass. It therefore follows that:

$$\sum_{i=1}^{N} g_i = 0 \tag{1.2}$$

Note that either some of the g_i are negative or all are zero. The key aspect of the method is the selection of k such that:

$$\sum_{i=1}^{k} g_i , \quad 1 \le k \le N$$

is minimum. Then the cells $a_1, a_2, \cdots, a_k$ are switched to the set B and the cells $b_1, b_2, \cdots, b_k$ are switched to the set A. Note that some of the individual interchanges may have actually *increased* the cut value, but that the *total* decrease in cut value is a maximum. As far as I know, this is the first algorithm applicable to the placement problem which permits uphill moves in an attempt to escape from a local minimum in the objective function.

Subsequent passes in the Kernighan-Lin procedure are performed until

---

[1] B. Kernighan and S. Lin, "An Efficient Procedure for Partitioning Graphs," (1970).

[2] D. Schweikert and B. Kernighan, "A Proper Model for the Partitioning of Electrical Circuits," (1972).

[3] M. A. Breuer, "Min-Cut Placement," (1977).

[4] A. Dunlop and B. Kernighan, "A Procedure for Placement of Standard-Cell VLSI Circuits," (1985).

[5] U. Lauther, "A Min-Cut Placement Algorithm for General Cell Assemblies Based on a Graph Representation," (1979).

$$\sum_{i=1}^{k} g_i \geq 0, \quad \forall k \in \{1, \ldots, N\}$$

At this point, a partition has been achieved such that the cut value is at a local minimum.

Schweikert and Kernighan extended the method to handle nets which connect to more than two cells.[1] Fiduccia and Mattheyses published a faster implementation of the Kernighan and Lin method.[2] Their method had a linear time complexity $O(P)$ with respect to the number of pins P, whereas the original Kernighan-Lin method was $O(N^2 \log N)$, where N is the number of cells. Each move in the Fiduccia-Mattheyses method is a single cell move, from one side of the partition to the other. Consequently, the extension to handling cells of various sizes is more straightforward than in the original Kernighan-Lin method. Dunlop and Kernighan remarked that their tests indicated that the quality of the solutions yielded by the Fiduccia and Mattheyses heuristic appeared not to be as good as those produced by the original method.[3] In any event, the time complexity of the Fiduccia and Mattheyses method appear to make it the more attractive approach as one extrapolates to the size of present day circuits and beyond.

Letting N_c be the number of cells and assuming for the moment that $N_c = 2^k$ where k is some integer, it is possible to define a placement procedure by recursively partitioning the cells into two equal halves, minimizing the cut values at each step. A wide variety of placement algorithms based on this *mincut* placement procedure have been published by Breuer.[4]

The main drawback with the recursive, or hierarchical, partitioning approaches proposed by Kernighan and Lin, as well as Schweikert and Kernighan, is that the partitioning of a block of cells at a particular level of hierarchy ignores the connections to cells in the other blocks. For example, in Figure 1.10a, a five-pin net interconnecting pins t_1, t_2, t_3, t_4 and t_5 is shown. The left (shaded) side of the top-level block has just been partitioned into two sets, L_1 and L_2. In the next step it is necessary to partition the right side into R_1 and R_2. Because the partitioning process ignores the connections to t_1 and t_2, the pins t_3, t_4 and t_5 can end up in partition R_2.

---

[1] D. Schweikert and B. Kernighan, "A Proper Model for the Partitioning of Electrical Circuits," (1972).

[2] C. Fiduccia and R. Mattheyses, "A Linear Time Heuristic for Improving Network Partitions," (1982).

[3] A. Dunlop and B. Kernighan, "A Procedure for Placement of Standard-Cell VLSI Circuits," (1985).

[4] M. A. Breuer, "Min-Cut Placement," (1977).

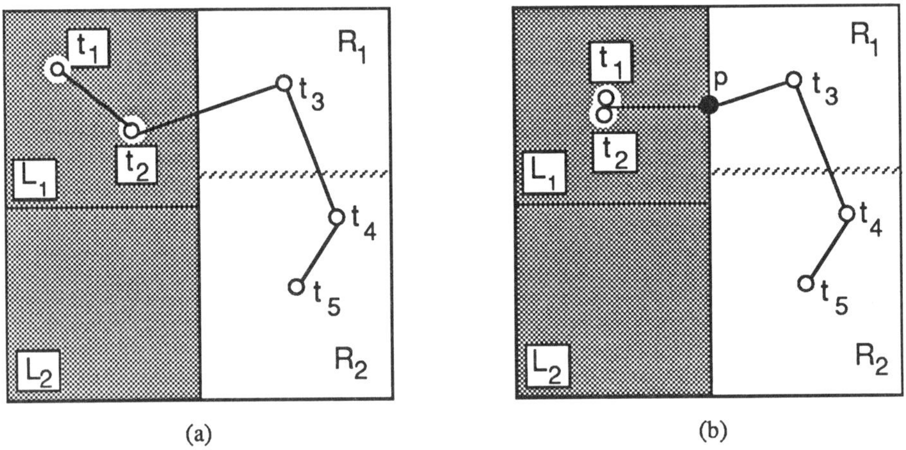

Figure 1.10 Terminal propagation.

Dunlop and Kernighan significantly improved the mincut placement technique by the method of *terminal propagation*.[1] Essentially, when partitioning a block of cells at some level of hierarchy, the connections from cells inside the block to cells outside the block are explicitly accounted for during the partitioning process. In Figure 1.10b, the pins t_1 and t_2 are assumed to lie in the center of their partition. Their location is *propagated* to point p on the boundary of partition R_1. The propagated pin (terminal) p participates in the partitioning process and thereby influences the ultimate locations of pins t_3, t_4 and t_5. Dunlop and Kernighan reported that terminal propagation makes a substantial difference in the quality of the solutions obtained, reducing the chip areas by an average of 30 percent. In order to use terminal propagation, the recursive partitioning process must proceed in a breadth-first manner, rather than depth-first.[2]

Lauther utilized the mincut algorithm in conjunction with a polar graph representation to yield a placement algorithm for cells of any rectangular size.[3] This method is extensively used for macro block placement.

A class of placement algorithms has been developed which ignores the particular restrictions of the image during the first stage.[4] [5] That is, the fact that each cell must be placed in a *slot* is temporarily ignored. With no restrictions as to where a cell may be placed on the image, the placement problem can be solved exactly in the sense that a global

[1] A. Dunlop and B. Kernighan, "A Procedure for Placement of Standard-Cell VLSI Circuits," (1985).

[2] A. Aho, J. Hopcroft, and J. Ullman, "The Design and Analysis of Computer Algorithms," (1974).

[3] U. Lauther, "A Min-Cut Placement Algorithm for General Cell Assemblies Based on a Graph Representation," (1979).

[4] C. K. Cheng and E. S. Kuh, "Module Placement Based on Resistive Network Optimization," (1984).

[5] J. P. Blanks, "Initial Placement of Gate Arrays Using Least-Squares Methods," (1984).

minimum can be achieved with respect to the sum of the squares of the net lengths.[1] These methods tend not to give good results, however, since the mapping of the cells back into legal slots tends to be a very difficult problem. A fairly complete bibliography of placement techniques was published recently.[2]

Prior to 1982, and to some extent a few years beyond that time, automatic placement algorithms were not widely used in industry. There were two primary reasons: (1) The existing algorithms simply did not achieve placements of the quality that could be achieved manually, albeit in less time. Essentially all of the algorithms attempted to minimize the total estimated routing length. This metric is fast to evaluate and its minimization can lead to reasonable chip areas and chip performance. However, the metric fails to address wireability and often fails to achieve the chip areas and chip performance expected by and obtainable by human designers. It was clearly necessary to find not only an algorithm capable of finding lower local minima in the total routing length objective function, but also a strategy which could more closely address chip area minimization as well as performance optimization.

(2) No less importantly, the algorithms consistently failed to capture all of the basic requirements and flexibility demanded by industrial layout problems. For example, row-based algorithms used for gate array and standard cell designs could not handle macro blocks on the same chip. For the macro/custom cell placement problem, some of the cells (custom cells) also required pin placement and/or aspect ratio selection as well as instance selection. These features rarely could be accommodated by existing placement algorithms.

1.3.2 Previous global routing methods

A good review of global routing strategies has been published.[3] For row-based images, to the best of my knowledge, there has been only one global routing model used. Figure 1.11 illustrates this model.

The dotted mesh grid overlays the entire core region. The vertical dotted edges are uniformly spaced, and similarly, the horizontal dotted edges are uniformly spaced. Each vertical edge, which crosses one row, is assigned a capacity which reflects the number of horizontal routing tracks which cross the vertical edge. Furthermore, each horizontal edge in the mesh grid is assigned a capacity which reflects the number of vertical routing tracks crossing the horizontal edge. If the row outlines are removed from Figure 1.11, and if a new graph is generated which resembles a rectangular dual of that in Figure 1.11, then the graph in Figure 1.12 is obtained.

Figure 1.12 is created by placing a node in the center of each rectangular region enclosed by the meshes of Figure 1.11, and by adding an edge between two adjacent nodes if there are routing tracks available across the perpendicular mesh edge between the two nodes. The capacity of each dark edge in Figure 1.12 represents the capacity of the mesh edge to which it corresponds. When a net is to be routed, each pin belonging to the net is

---

[1] C. K. Cheng and E. S. Kuh, "Module Placement Based on Resistive Network Optimization," (1984).

[2] T. Hildebrandt, "An Annotated Placement Bibliography," (1985).

[3] E. S. Kuh and M. Marek-Sadowska, "Global Routing," (1986).

assigned to the nearest node in the graph of the type shown in Figure 1.12. For example, a three-pin net, in which the pins are assigned to the nodes labeled A, B, and C, is routed as shown in Figure 1.12.

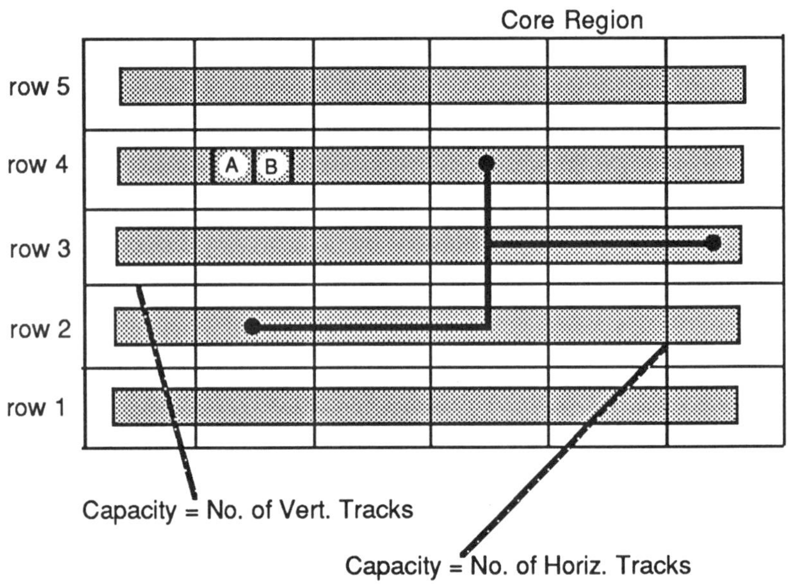

Figure 1.11 A gate array global routing model.

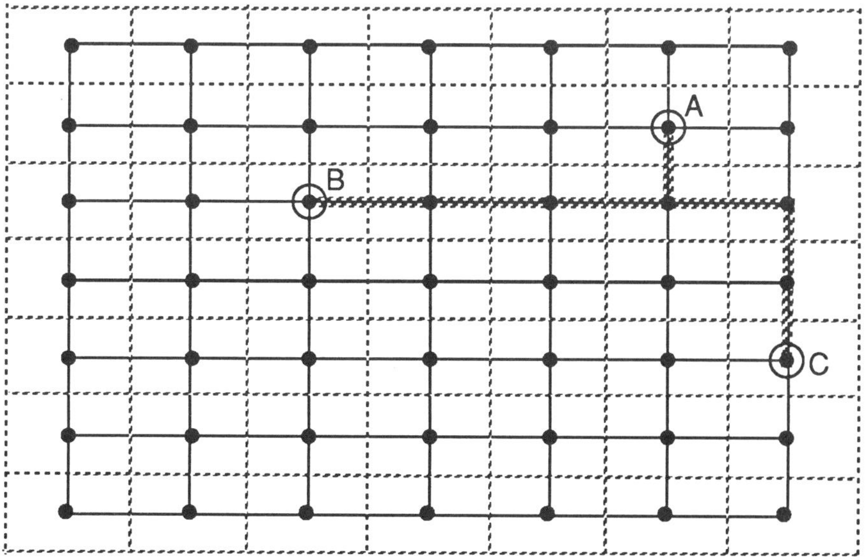

Figure 1.12 Routing graph generated from the mesh grid of Figure 1.11.

The basic problem confronting a global router may be stated as follows: Minimize the sum of the routing lengths of all of the nets subject to the satisfaction of the *capacity constraints* of the edges. The constraints result from the fixed widths of the channel edges. The most common approach to global routing has been sequential routing. That is, one net is routed at a time. With this approach, the shortest possible route is selected whenever channel-edge capacities permit it.

It is generally not possible to route all nets with the shortest possible interconnection paths because the nets compete with each other for the channel edges to be used. Two of the more common approaches directed toward the solution to this problem are: (1) Route all nets independently, and if some channel edges become excessively congested, reroute the nets causing the excessive congestion.[1] (2) The second approach is to route nets sequentially, and after each net is routed, update the information about channel-edge congestion. Fully utilized channel edges are no longer available for the routing of future nets. Obviously, the routing order becomes a crucial factor. Arguments can be made for routing the shortest nets first, and also for routing the nets with the most pins first. It appears impossible to devise a general procedure which works the best in all situations.

For a multi-pin net on a graph, the shortest path problem is the well-known graph-based Steiner tree problem.[2] For arbitrary numbers of pins per net, this problem belongs to the class of NP-complete problems. However, exact solutions to this problem are

[1] R. Prim, "Shortest Connecting Networks and Some Generalizations," (1957).

[2] S. Hakimi, "Steiner's Problem in Graphs and its Implications," (1971).

known to exist for two-pins nets[1] and three-pin nets.[2] N. P. Chen describes two heuristics for handling nets containing more than three pins.[3]

Most of the previous approaches which avoid the net routing-order dependence problem restrict the net list to contain only two-pin nets. Also, the shapes of the routes are limited to a few feasible patterns. Usually only one or two bends per net are allowed. With these restrictions, a successful 0-1 linear programming approach has been reported.[4]

Recently, an integer linear programming approach has been applied to the above global routing problem.[5] [6] [7] The advantage of this formulation is that the problem instance is viewed in its entirety, and does not depend on any sequential ordering of the nets. An efficient procedure was developed for approximately solving such an integer program. Furthermore, the approximation is provably good in the sense that the solution is guaranteed to be close to the best possible. Finally, the procedure runs in polynomial time. A drawback to this approach is that multi-pin nets are not always handled correctly. For example, the method can produce *track sharing* or *routing cycle* problems for multi-pin nets.[8]

A newer method, also based on integer linear programming, has been developed with multi-pin nets in mind.[9] However, while the method appears to work well for three-pin nets, it may not be practical for nets with more than three pins. In practice, the two-pins per net restriction and even a three-pins per nets limit are very unrealistic, and the search has continued for good heuristics to handle multi-pin nets, which usually comprise a significant portion of the nets.

Accuracy can be a serious limitation with the mesh grid approach. Typically, a vertical mesh grid segment, such as that shown in Figure 1.13, encompasses routing tracks from two or more adjacent channels. Mesh segment *MG1* encompasses routing tracks from both channel 3 and channel 2. Similarly, grid *MG2* includes routing tracks from channel 2 and channel 1. Consider the situation illustrated in Figure 1.13 in which a vast majority of the nets crossing a grid such as *MG1* are in only one of the channels, for example, channel 2. The nets designated as *B* clearly outnumber those designated as *A*. A

---

[1] E. Dijkstra, "A Note on Two Problems in Connection with Graphs," (1959).

[2] N. P. Chen, "New Algorithms for Steiner Tree on Graphs," (1983).

[3] N. P. Chen, "New Algorithms for Steiner Tree on Graphs," (1983).

[4] R. M. Karp, et. al., "Global Wire Routing in Two Dimensional Arrays," (1983).

[5] A. Ng, P. Raghavan, and C. Thompson, "Specification Languages for Describing Rectilinear Steiner Tree Configurations," (1986).

[6] P. Raghavan and C. Thompson, "Provably Good Routing in Graphs: Regular Arrays," (1985).

[7] P. Raghavan and C. Thompson, "Multiterminal Global Routing: A Deterministic Approximation Scheme," (1987).

[8] A. Ng, P. Raghavan, and C. Thompson, "Experimental Results for a Linear Program Global Router," (1987).

[9] A. Ng, P. Raghavan and C. Thompson, "Experimental Results for a Linear Program Global Router," (1987).

similar situation is present in grid *MG2* just below *MG1*. Most of the nets (*C*) cross in channel 2 as opposed to channel 1 (*D*). Even though the total capacity is not exceeded for either *MG1* or *MG2*, the actual capacity of channel 2 can be significantly exceeded, leading to an unroutable chip.

Another problem which often arises is also shown in Figure 1.13. The routes designated as *E* and *F* in channel 2 are within the capacity limitations of their corresponding mesh grids. However, between the two mesh grids, the routes designated as *G* significantly exceed the capacity of channel 2. This again yields an unroutable chip. This problem can be controlled by putting artificially low capacities on the mesh grids, to decrease the probability of the capacity being exceeded between the grids. However, this has the unpleasant side effect of forcing subsequent routes to take long detours around regions which are reported to be blocked, but which are in fact not excessively congested.

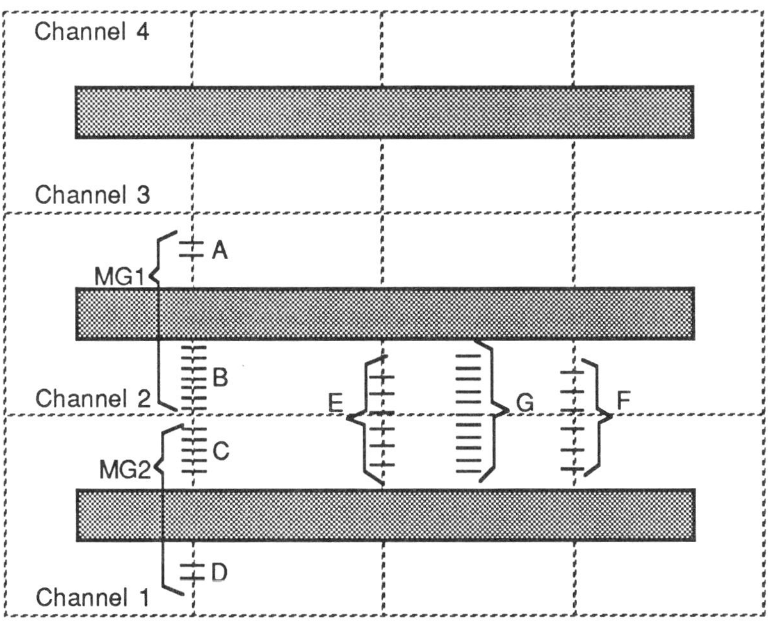

Figure 1.13 Accuracy limitations arise if mesh segments encompass more than one channel, or if they are excessively spaced.

The problems described in the two previous paragraphs can both be controlled to some extent by using a much finer gridspacing. For example, by using one or more horizontal grids per channel and a very close spacing between the vertical grids. This can be unacceptable for large circuits since the run time of the global routers is usually a strong function of the number of mesh rectangles. The run time can easily be $O(n^3)$ or more, where n is the number of mesh rectangles.[1] Also, the additional horizontal grids give rise to another problem. Since each pin invariably has one or more electrically equivalent pins, having the two sides of a row belong to different horizontal slices (containing vertical cut lines) means that an equivalent set of pins will belong to different horizontal slices. In order to compute accurate crossings, the global router must account for and take advantage of electrically equivalent pins. To the best of my knowledge, existing algorithms have not done so. It will be shown in Chapter 3 that if the electrically equivalent pins are efficiently utilized, then up to 20 percent of the chip area reserved for routing can be eliminated.

Global routing for the macro/custom cell case is similar to the mesh grid problem, except that the graph derived from the mesh grid is replaced by a channel graph. An example is shown in Figure 1.14. The routing order of the nets is the crucial problem here. Those methods which avoid the routing order dependence fail to allow multi-pin nets, which is unrealistic in industrial circuits. Also, existing methods fail to optimize the routing length of nets in the presence of electrically equivalent pins.

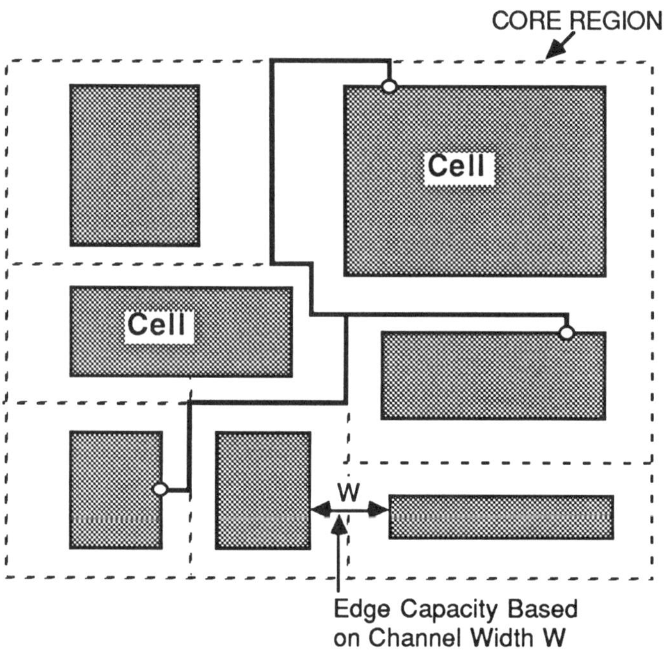

Figure 1.14 Macro/custom cell global routing model.

[1] N. P. Chen, "New Algorithms for Steiner Tree on Graphs," (1983).

1.4 A New Approach to Cell-Based Placement and Global Routing

In 1983, Kirkpatrick, Gelatt, and Vecchi introduced a new general combinatorial optimization technique known as simulated annealing.[1] They proposed that simulated annealing was an effective method for solving combinatorial optimization problems involving many degrees of freedom. The basic feature of simulated annealing is that the configuration space of the optimization problem is explored while permitting *hill climbing* moves, that is, the acceptance of new configurations of the problem which increase the cost. These uphill moves are controlled by a parameter which is analogous to temperature in the physical annealing process, and are less and less likely towards the end of the process. The concept of permitting uphill moves during intermediate stages of the optimization process was actually used by Kernighan and Lin in their highly-successful partitioning algorithm.[2] To the best of my knowledge, the first report of a macro cell placement program using simulated annealing was by Jepsen and Gelatt in 1983.[3] Subsequently, in 1984, the TimberWolf placement and routing package was introduced, which presented implementations of simulated annealing for gate array, standard cell, and macro/custom cell placement and global routing.[4] An implementation of simulated annealing for floorplan design was reported by Otten and van Ginneken, also in 1984.[5] More recently, another implementation of simulated annealing was reported by Wong and Liu for floorplan design.[6] An overview of the simulated annealing algorithm is presented in Chapter 2.

The basic simulated annealing algorithm is very easily stated and it is quite simple to obtain an implementation for many combinatorial optimization problems. In many cases, it is possible to obtain a good solution by means of simulated annealing algorithm for a problem in which it is very difficult to develop a direct, or non-probabilistic, heuristic due to very many degrees of freedom. However, the quality of the solutions obtained by simulated annealing can vary dramatically from one implementation to another. Unfortunately, much experimentation is usually required to develop a high performance simulated annealing algorithm for a new application. This is due to the lack of definitive theoretical results concerning guidelines for optimal properties and control of a simulated annealing algorithm.

The two main characteristics of a simulated annealing algorithm are: (1) a cost function reflecting the quality of the current state, and (2) a new-state generation function allowing the exploration of the configuration space. Such simplicity permits the addition of a great number of problem-specific constraints into the cost function and the generation of new states function. For example, in a placement problem, non-rectangular cells, uniform-height cells with widely varying widths, critical-net weighting, cell exchange classes, and

---

[1] S. Kirkpatrick, C. Gelatt, and M. Vecchi, "Optimization by Simulated Annealing," (1983).

[2] B. Kernighan and S. Lin, "An Efficient Procedure for Partitioning Graphs," (1970).

[3] D. Jepsen and C. Gelatt, "Macro Placement by Monte Carlo Annealing," (1983).

[4] C. Sechen and A. Sangiovanni-Vincentelli, "The TimberWolf Placement and Routing Package," (1984).

[5] R. Otten and L. van Ginneken, "Floorplan Design using Simulated Annealing," (1984).

[6] D. Wong and C. Liu, "A New Algorithm for Floorplan Design," (1986).

exact pin locations can all be accommodated with ease. Furthermore, with respect to global routing, it is possible to consider exact net embeddings and also to eliminate the routing-order dependence problem.

Even while incorporating tremendous flexibility, the simulated annealing algorithm achieves excellent optimization results due to the controlled acceptance of uphill moves. The main focus of this work was to demonstrate that it was possible to devise placement and global routing algorithms which capture all of the flexibility required by state-of-the-art industrial circuits. Less attention was to paid to minimizing the computation time.

Chapter 3 presents the algorithms and the implementation of a new standard cell placement and global routing package. The package is also applicable to the row-based gate array placement and global routing problem. This package, named *TimberWolfSC*, makes extensive use of the simulated annealing algorithm. The cells are permitted to have any shape or size. Macro blocks are also permitted on the chip, and pad placement along the periphery of the core region is also performed. The placement of the cells is first optimized with respect to total interconnect length. Since very many new states are considered during the simulated annealing process, it is necessary to choose an objective function which requires a minimum of computation time to update its value. The total interconnect length objective function offers the best compromise of speed versus chip area and circuit performance optimization. After the minimization of the total interconnect length, TimberWolfSC enters a placement-refinement stage, in which local changes are made to the placement should such a change lead to a reduction in the number of wiring tracks required. The computation of the total number of wiring tracks requires a global routing algorithm. A new algorithm was developed, based on the *total channel density* model for row-based layout images. This model is illustrated in Figure 1.15.

All pins, including electrically equivalent pins, are considered during the global routing process. At any step in the optimization, a feasible global routing for the entire chip is maintained. The key aspect of the new global router is the optimization of the placement of *switchable net segments*. For example, in Figure 1.15, a net consisting of seven electrically equivalent pin pairs is shown. The net segment which must interconnect pin pairs P_2 and P_3 can be implemented as either the net segment A_1 or the net segment A_2. A similar situation exists for the interconnection of pin pairs P_5 and P_6, as well as P_7 and P_8.

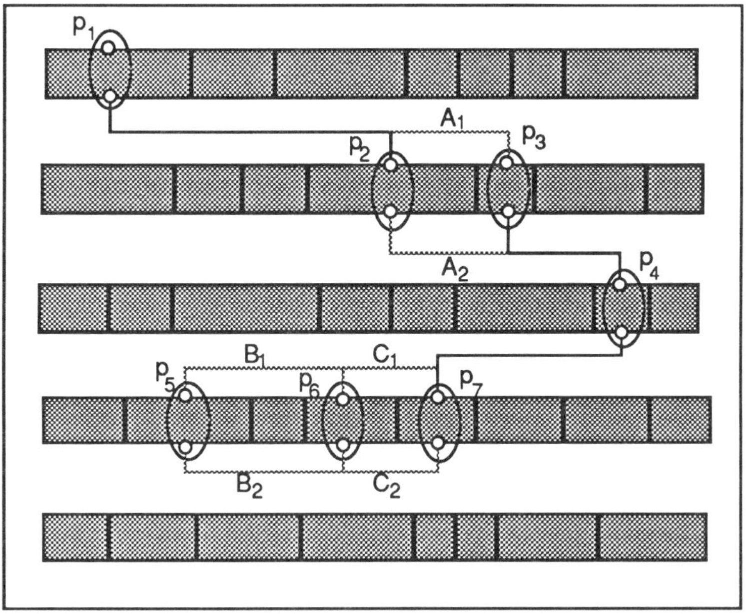

Figure 1.15 Total channel density model for row-based layout images.

Results on industrial circuits showed that TimberWolfSC yielded area savings ranging from 15 to 75 percent versus existing automatic and manual layout methods. For all circuits tested, the global router reduced the number of wiring tracks needed by an additional 7 to 16 percent in comparison to total interconnect length minimization alone. Furthermore, the placement-refinement stage produced wiring-track reductions ranging from 15 to 25 percent in comparison to total interconnect length minimization alone. TimberWolfSC also features critical-net weighting for performance-driven placement.

Chapter 4 presents the algorithms and the implementation of a new macro/custom cell placement and global routing package. This package, named *TimberWolfMC*, makes extensive use of the simulated annealing algorithm. TimberWolfMC is applicable to circuits containing cells of any rectilinear shape. Furthermore, the cells may have fixed geometry including pin locations (macro cells) or the cells may have an estimated area with a specified aspect ratio range, and with pins that need to be placed (custom cells). The pins for a custom cell may be specified in several possible ways: (1) a pin may be given a particular fixed location, (2) a pin may be assigned to a particular edge or edges of a cell, (3) aa pin may belong to a group of pins which may be assigned to a particular edge or edges of a cell, or (4) a pin may belong to a group of pins which is assigned a particular sequential ordering as well as a particular edge or edges of a cell. The placement of a single pin, a group of pins, or a sequence of pins may be specified as being restricted to either one cell edge, two cell edges, or any of the edges.

TimberWolfMC also permits the custom cells to have aspect ratios in a continuous or discrete range. Furthermore, several alternative instances for a custom cell may be specified, whereby TimberWolfMC is to select the one which is most suitable. The selection of aspect ratio and/or instance is guided by the minimization of the total interconnect length and by the geometry of the empty space allotted for the cell as influenced by the neighboring cells. TimberWolfMC places circuits consisting entirely of macro cells as well as circuits consisting entirely of custom cells. Furthermore, the program will place circuits consisting of a combination of macro and custom cells. Consequently, TimberWolfMC is applicable to chip planning problems. The total interconnect length calculation is based on the exact pin locations (as opposed to assuming that all of the pins associated with a cell have a location equal to the center of the cell). Consequently, all eight possible orientations are considered for each cell.

An algorithm for accurately estimating the total interconnect length, for both random and optimized placements is introduced in Chapter 5. The algorithm developed in Chapter 6 for accurately estimating interconnect *area* requires a good estimate of the *final* total interconnect length. This estimate is needed *before* the macro/custom cell placement algorithm is executed. This placement algorithm is based on the simulated annealing algorithm. A characteristic feature of simulated annealing is long computation times. Hence the possibility of executing the placement algorithm twice, once for ascertaining the final total interconnect length and the second for obtaining the final placement using the interconnect-area estimator, is unreasonable. An estimate of the average total interconnect length for random placements is not difficult to obtain. The placement algorithm can explore a statistically meaningful number of random placements and compute the average total interconnect length. Importantly, this estimate can be achieved in negligible computation time.

Assume that it is possible to calculate the ratio (R_{opt}) of the total interconnect length for random placement over optimized placement. By measuring the random interconnect length (L_R) for a circuit, the final total interconnect length can be estimated by

$$L_{opt} = \frac{L_R}{R_{opt}} \tag{1.3}$$

The algorithms developed in Chapter 5 enable the use of Eqn. 1.3 to compute an accurate estimate of the final total interconnect length for optimized placements.

An algorithm for accurately estimating the interconnect area around the individual cells, introduced in Chapter 6, is used in TimberWolfMC. This interconnect-area estimation algorithm has resulted in the generation of placements which require virtually no placement modification during or after detailed routing.

The placement algorithm proceeds in two distinct stages. During the first stage, the interconnect area around the individual cells is determined using the *dynamic* interconnect-area estimator. That is, as each cell is moved, its effective area is adjusted. A simulated annealing algorithm is used to minimize the total interconnect length. The second stage of TimberWolfMC consists of several executions of a placement-refinement algorithm. Each execution consists of three steps: (1) a channel definition step, (2) a global routing step, and (3) a placement-refinement step. The information obtained in step 2 is used to compute the density of all of the channels, which then permits accurate interconnect-area determination. The placement of the cells is then refined in step 3 to reflect the required interconnect area. A low-temperature simulated annealing algorithm is used.

Total interconnect length savings of 8 to 49 percent were achieved in experiments on 9 industrial circuits. Furthermore, for the 9 examples, circuit-area reductions ranged from 4 to 56 percent versus a variety of other placement algorithms.

Chapter 7 describes an edge-based channel definition algorithm for circuits consisting of cells of any rectilinear shape. This algorithm is particularly well suited for channel definition just prior to placement refinement. Furthermore, the algorithm may be employed as a basic channel definer just prior to the execution of global and then detailed routers.

Automatic layout of a macro/custom cell circuit begins with the total interconnect length minimization stage. At this point, four basic steps remain to be completed. (1) A placement-refinement step is performed in which any (usually small) inaccuracies produced by the dynamic interconnect area estimator are corrected. That is, if insufficient space was allocated between a pair of cells, then additional space is provided as required. Or, if excessive space was allocated, then the cells are compacted as much as possible. (2) The areas (channels) available for routing are identified. This is known as the *channel definition* step. (3) The nets (as specified by the net list) are *globally* routed. By taking a global view of the placement of the interconnections, the global router assigns each portion (segment) of a net to a specific channel. (4) Interconnect instances are created for each of the net segments in each channel. That is, a *detailed router* is used to give each net segment a specific size, shape and location.

In order to perform the placement-refinement step (step 1 above), it is necessary to use steps 2 and 3. The same step 3 global router may be used in step 1. However, a channel definition algorithm applicable to step 2 may be unsuitable for use in step 1.

The channel definition algorithm described in Chapter 7 may be used either in the placement-refinement phase or prior to the detailed routing phase in the automatic layout procedure. This channel definition algorithm generates a definitive routing order for the channels. The method assumes that a generalized channel router is available, that is, a router which can handle channels of $n \geq 4$ sides, with fixed pins on any of the n sides. Such a generalized channel router is currently available.[1] The algorithm which prescribes the routing order of the merged channels is a generalization of the slicing structure for the arbitrary rectilinear cell case.[2] [3]

The global routing algorithm used in TimberWolfMC is introduced in Chapter 8. This algorithm has the following features: (1) It is a general-purpose global router which avoids the classical routing-order dependence problem. (2) Multi-pin nets are handled in the same fashion as two-pin nets. (3) The global router fully utilizes electrically equivalent pins to minimize the routing length of a net. (4) The global router is independent of the layout style since the only inputs to the algorithm are a net list and a channel graph. (5)

---

[1] H. Shin and A. Sangiovanni-Vincentelli, "A Detailed Router Based on Incremental Routing Modifications: MIGHTY," (1987).

[2] R. H. Otten, "Automatic Floorplan Design," (1982).

[3] A. Sangiovanni-Vincentelli, Private Communication, (1986).

The global router minimizes the sum of the routing lengths of all of the nets subject to the satisfaction of the *capacity constraints* of the edges.

The original program in the TimberWolf package was a generalized *island-style* gate array placement program. This program, presented in the Appendix, served to indicate the inherent flexibility of the simulated annealing algorithm.

An island-style gate array is such that an array of cells covers the chip with pads on the periphery. Gate array designs often allow additional flexibility and hence greater gate utilization by creating functionally independent units within a cell. For example, some gate arrays utilize functional units which are half-cell sized. TimberWolf handles functional units which are full-cell, half-cell, or quarter-cell sized. The term *module* is used to refer to a fundamental unit specified in the net list. A module may be the size of: (1) a full cell, (2) a half cell, or (3) a quarter cell. A *macro module* consists of a prewired collection of cells. TimberWolf permits these macro modules to be arbitrarily shaped.

Additional flexibility was introduced by defining *exchange classes* of modules. The modules in an exchange class are only allowed to be interchanged with other modules belonging to the same class. Another feature is the ability to restrict the permissible locations for macro modules. For example, it is often the case that gate arrays have wider channels in the center of the array. This is in anticipation of the greatest wiring congestion occurring in this region. Because prewired macro modules usually have fixed cell-to-cell spacing, certain macros may not be placed in the center region (or, on the other hand, the outer regions). TimberWolf allows the designation of cell locations as either suitable or unsuitable for a particular set of macro modules.

The TimberWolf gate array placement program was tested on several standard benchmark placement problems. The placements generated by TimberWolf had final total interconnect length values which were 6 to 27 percent smaller than the previously published results.

Chapter 2

The Simulated Annealing Algorithm

2.1 Introduction

Most placement and global routing problems belong to the class of NP-complete problems.[1] For problems in this class, there is no known exact algorithm whose worst-case time complexity is bounded by a polynomial in the size of the input. Consequently, heuristic algorithms are used to find solutions to these problems. These algorithms explore a discrete space of admissible configurations S in a deterministic fashion. Starting from an initial configuration $i_0 \in S$, a sequence of configurations are explored until a satisfactory one is found. The algorithm is specified by the rules according to which configurations are generated and accepted. Often the search terminates in a local minimum, at a state $\hat{j} \in S$ such that $c(j) \geq c(\hat{j})$ for all $j \in \hat{S}(\hat{j})$ where $c()$ is the cost function defined on the configuration space and where $\hat{S}(\hat{j})$ represents the states of S reachable from $\hat{j}$. The state $\hat{j}$ is a local minimum if there exists a state $\bar{j} \in S$ such that $c(\bar{j}) < c(\hat{j})$. In many cases, the quality of the local minimum states, or configurations, are quite inferior in comparison to the global minimum state. Heuristic algorithms often end at these local minimum states because their behavior is essentially *greedy*, that is, they only accept new states which maximally reduce the cost.

---

[1] M. Garey and D. Johnson, "Computers and Intractability: A Guide to the Theory of NP-Completeness," (1979).

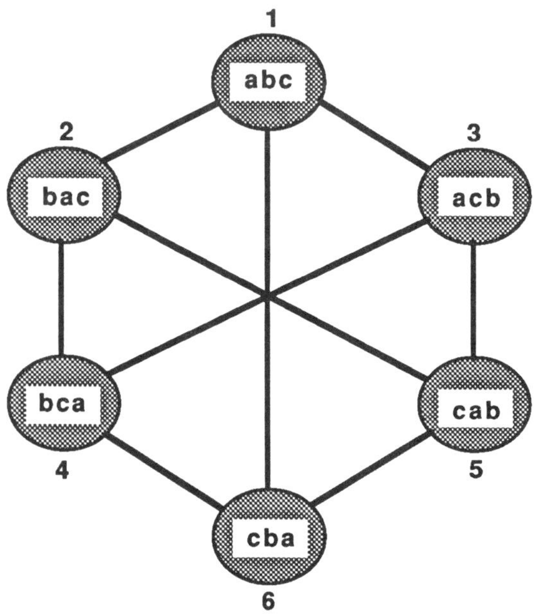

Figure 2.1 State graph representing all permutations of the set $\{a, b, c\}$.

Consider the problem illustrated in Figure 2.1. Each state is represented by a permutation of the set $\{a, b, c\}$. Each state is represented by a node in the state graph of Figure 2.1. For any given state, the only reachable new states are those which can be obtained by swapping the positions of two of the elements of the set. Two states are interconnected by an edge in the state graph if one state may be generated from the other by a *legal* swap. Suppose it is given, as indicated in Figure 2.2, that the cost of each state is such that:

$$c(1) < c(5) < c(3) < c(6) < c(2) < c(4)$$

A directed edge from state i to state j indicates that state j can be generated from state i, and that the cost of state j is lower than that of state i. A greedy heuristic search algorithm, beginning in state 6, may well advance to state 5 by the interchange of elements b and a. However, once in state 5, the greedy algorithm has achieved a local minimum whose cost is greater than the global minimum, represented by state 1.

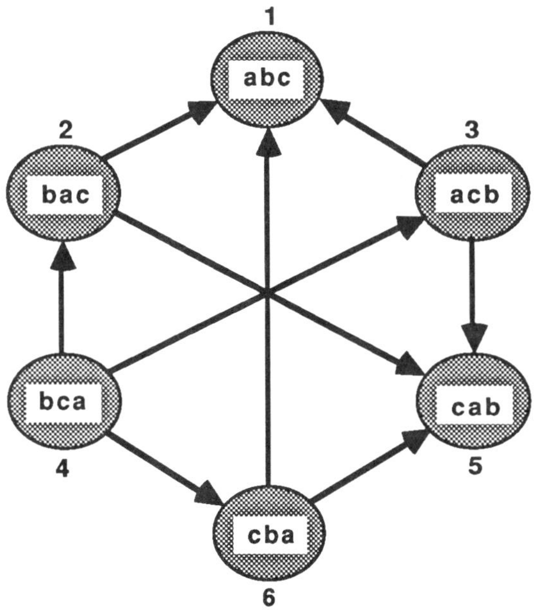

Figure 2.2 State graph in which directed edges point to lower cost states.

A class of randomized algorithms have been developed which seek to avoid this behavior.[1] These algorithms generate the next new state randomly and are able to *climb hills*, that is, to generate and accept new states which have higher cost than the current state. Simulated annealing, as proposed by Kirkpatrick, Gelatt, and Vecchi, also permits *hill climbing* moves.[2] However, as opposed to other randomized algorithms, the generation and acceptance of these moves are controlled by a specific criterion based on the current state of the search process. The controlling mechanism is based on the observation that combinatorial optimization problems with a large configuration space exhibit properties resembling those of physical processes with many degrees of freedom. The process of cooling a fluid into a low energy state, for example, when growing a crystal, was shown to be a process similar in nature to that of finding an optimum solution of a large combinatorial optimization problem.[3] With respect to growing crystalline solids, physical annealing has long been used and is well understood. The basic process begins with raising the fluid, or *melt*, to a very high temperature. Subsequently, the temperature is slowly lowered until the crystal is formed. At the point at which the material begins to freeze, the temperature must be very slowly reduced. The Metropolis Monte Carlo method can be used to simulate the

[1] J. T. Schwartz, "Fast Probabilistic Algorithms for Verification of Polynomial Identities," (1980).

[2] S. Kirkpatrick, C. Gelatt, and M. Vecchi, "Optimization by Simulated Annealing," (1983).

[3] S. Kirkpatrick, C. Gelatt, and M. Vecchi, "Optimization by Simulated Annealing," (1983).

process.[1 2] It has been proposed as an effective method for obtaining good solutions to combinatorial optimization problems.[3]

With respect to combinatorial optimization, the Metropolis method begins with an arbitrary initial configuration. Subsequently, given that the state of the simulation is represented by configuration i at time m where $m = 0, 1, 2, \cdots$, a new configuration j is randomly generated from an admissible set of neighboring configurations, or states. The cost of the proposed new state is then subjected to the acceptance criterion which, based on the value T_m of a controlling parameter termed *temperature* at time m, determines whether the new state is to be accepted. If the cost is not increased, the simulation always accepts the move. On the other hand, if the cost increases, a random real number r is drawn from a uniform distribution of reals in the interval [0, 1]. Then, another real number y is computed by:

$$y = \exp\left\{-\frac{c(j) - c(i)}{T_m}\right\}$$

where $c()$ is the cost function defined on the configuration space. If $r \leq y$ the simulation accepts the move; otherwise, the move is rejected. In any event, the time m is incremented. Note that the higher the temperature is, the higher the probability of accepting a hill-climbing move. The initial temperature, the number of moves generated at each temperature and the rate of decrease of temperature are all important parameters that affect the speed of the algorithm and the quality of the final configuration.

With respect to the example of Figure 2.2, the fact that simulated annealing allows hill-climbing moves leads to the generation of the graph shown in Figure 2.3. Here the dotted edges indicate moves which increase the cost, while the solid edges represent moves which decrease the cost. Should the algorithm find itself in state 5, there is a nonzero probability that it will advance to state 3. From state 3, there exists a down hill edge to state 1, the global minimum state.

The remainder of this chapter is organized as follows: Section 2.2 presents the basic simulated annealing algorithm. Some results concerning the theoretical investigations of the simulated annealing algorithm are covered in Section 2.3. Section 2.4 overviews work on general annealing schedules. In Section 2.5, I present details shared by implementations of simulated annealing for placement and global routing. A comparative study of two different new state acceptance functions is described in Section 2.6. Finally, in Section 2.7, I present a method for the fast evaluation of the exponential function, an operation which is usually performed very many times during the execution of a simulated annealing algorithm.

---

[1] N. Metropolis, A. Rosenbluth, M. Rosenbluth, and A. Teller, (1953).

[2] K. Binder, "Monte Carlo Methods in Statistical Physics," (1978).

[3] S. Kirkpatrick, C. Gelatt, and M. Vecchi, "Optimization by Simulated Annealing," (1983).

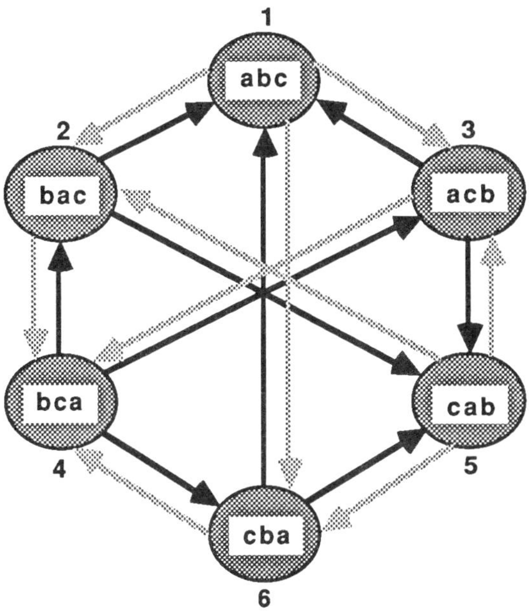

Figure 2.3 State graph with uphill moves indicated by dotted edges.

2.2 The Basic Simulated Annealing Algorithm

The following function gives the general structure of the class of algorithms called *probabilistic hill-climbing algorithms,* of which simulated annealing is a special case. This class was proposed by Romeo and Sangiovanni-Vincentelli, where a number of different algorithms with the same structure were introduced.[1]

---

[1] F. Romeo and A. Sangiovanni-Vincentelli, "Probabilistic Hill Climbing Algorithms: Properties and Applications," (1985).

```
Sim_Anneal ( i0 , T0 ) {
        T = T0 ;
        i = i0 ;
        while( stopping criterion is not satisfied ) {
                while( inner loop criterion is not satisfied ) {
                        j = generate( i ) ;
                        if( accept( c( j ), c( i ), T ) ) {
                                i = j ;
                        }
                }
                T = update( T ) ;
        }
}
```

The acceptance of a new state j is determined by accept(), whose structure is shown below.

```
accept ( c( j ) , c( i ) , T ) {
        Δc = c( j ) − c( i );
        y = f( Δc , T ) ;
        r = random( 0 , 1 ) ;
                /*
                 *      random( 0, 1 ) returns a random number in the
                 *      interval [0, 1] (with uniform distribution).
                 */
        if( r < y ) {
                return( 1 ) ;
        } else {
                return( 0 ) ;
        }
}
```

The general form of the simulated annealing algorithm, as presented above, is characterized by: (1) the generation function generate(), (2) the acceptance function accept(), (3) the updating function update(), (4) the *inner loop criterion* and (5) the *stopping criterion.*

Kirkpatrick, Gelatt, and Vecchi used the following formulation of the acceptance function f( ) :[1]

$$\mathbf{f}\,(\Delta c\,,T) = \min\left\{\,1.0\,,\ \exp\left(-\frac{\Delta c}{T}\right)\right\} \tag{2.1}$$

[1] S. Kirkpatrick, C. Gelatt, and M. Vecchi, "Optimization by Simulated Annealing," (1983).

In particular, the implementation of f() uses a Boltzmann-like factor. Note that when Δc is negative or zero, that is, when the cost of the new state is less than or equal to the cost of the current state, then the new state is always accepted. On the other hand, when Δc is positive, the acceptance probabilities are distributed according to the Boltzmann factor.

It is possible to vary the shape of f() by adjusting the control parameter T, called temperature. An updating rule for T proposed by Kirkpatrick, Gelatt, and Vecchi is given below:[1]

$$T_n = T \cdot \alpha(T) , \qquad 0 < \alpha(T) < 1 \tag{2.2}$$

where T_n is the new value and T is the current value of the temperature parameter.

In the function accept(), note that new states characterized by $\Delta c \leq 0$ always satisfy the acceptance criterion. However, for the new states characterized by $\Delta c > 0$, the parameter T plays a fundamental role. If T is very large, then r is likely to be less than y and a new state is almost always accepted irrespective of Δc. If T is small, close to 0, then only new states that are characterized by very small $\Delta c > 0$ have any chance of being accepted. In general, all states with $\Delta c > 0$ have smaller chances of satisfying the test for smaller values of T.

2.3 Theoretical Investigations of the Simulated Annealing Algorithm

Simulated annealing, as applied to combinatorial optimization problems, can be represented by Markov chains.[2 3 4] The analysis in this section follows that of Mitra, Romeo, and Sangiovanni-Vincentelli.[5] For illustrative purposes, a Markov chain model will be developed for the linear placement problem. Suppose that three interconnected modules $\{a, b, c\}$ must be placed on a one-dimensional grid such that the total interconnection length is minimized. The state space S consists of six (3!) configurations which represent all possible placements of the three modules, that is, $S = \{1, 2, 3, 4, 5, 6\}$ where

$$1 = \{\ a, b, c\ \}$$
$$2 = \{\ c, a, b\ \}$$
$$3 = \{\ b, c, a\ \}$$
$$4 = \{\ a, c, b\ \}$$
$$5 = \{\ c, b, a\ \}$$
$$6 = \{\ b, a, c\ \}$$

Assume that the simulated annealing algorithm applied to this problem generates new configurations by interchanging the positions of two elements. Define $N(i)$ to be the

---

[1] S. Kirkpatrick, C. Gelatt, and M. Vecchi, "Optimization by Simulated Annealing," (1983).

[2] S. Karlin, "A First Course in Stochastic Processes," (1973).

[3] W. Feller, "An Introduction to Probability Theory and Applications," (1970).

[4] D. Freedman, "Markov Chains," (1971).

[5] D. Mitra, R. Romeo, and A. Sangiovanni-Vincentelli, "Convergence and Finite-Time Behavior of Simulated Annealing," (1985).

configurations reachable from state i. That is, $N(i)$ are the states *neighboring* state i. For example, $N(1) = \{4, 5, 6\}$. The graph shown in Figure 2.1 is a schematic representation of the generation rule. Each node represents one of the configurations. There exists an edge between nodes i and j in the graph if j can be obtained from i by interchanging the positions of the two modules.

As indicated in Figure 2.2, suppose it is given that the cost of each state is such that:

$$c(1) < c(5) < c(3) < c(6) < c(2) < c(4)$$

The fact that simulated annealing allows hill-climbing moves leads to the generation of the graph shown in Figure 2.3. For the simulated annealing algorithm, the transition probability is appended to each edge in the graph. The solid edges in Figure 2.3 indicate moves which lower the cost. Hence, for these edges, the transition probability is independent of T. The dotted edges represent moves which result in an increase in cost. These are *hill climbing* transitions, whose transition probabilities are dependent on T. In general, the application of simulated annealing to a combinatorial optimization problem can be represented by a graph whose nodes are configurations, whose edges represent transitions which can be obtained by the generation rule, and whose edge labels represent the probability that the corresponding transition is generated and accepted by the algorithm.

Given that the configuration after the k-th iteration was i, the probability that the configuration selected by a simulated annealing algorithm is j at iteration $(k+1)$ is determined by

$$\text{Prob}\ \{\ X_{k+1} = j \ \mid\ X_k = i\ \} \equiv P_{ij}(T) \tag{2.3}$$

where X_k is the value of the random variable X representing the generic state of the algorithm, at the k-th iteration. In simulated annealing,

$$P_{ij}(T) = G_{ij}(T)\ \mathrm{f}(\Delta c_{ij}, T)\quad \forall j \in N(i) \tag{2.4}$$

where $G_{ij}(T)$ is the probability of generating state j given that the current state is i and given the current value of the parameter T. Furthermore,

$$\Delta c_{ij} = c(j) - c(i)$$

Note that $G_{ij}(T) = 0$ for the states which are not in $N(i)$. Since $G_{ij}(T)$ is a probability,

$$\sum_{j \in N(i)} G_{ij}(T) = 1.0 \tag{2.5}$$

Note that Eqn. 2.4 is the product of two different terms: $G_{ij}(T)$ is the probability that j is generated by the algorithm and f() is the probability that the new configuration will be accepted. An example of $G_{ij}(T)$ is given by

$$G_{ij}(T) = \begin{cases} \dfrac{1}{|N(i)|} & \forall j \in N(i) \\ 0 & \forall j \notin N(i) \end{cases} \tag{2.6}$$

Note that there is an equal probability of generating any state and that this probability is independent of T. For the previous small example, since $|N(i)| = 3$ for all $i \in S$,

$$G_{ij}(T) = \begin{cases} \frac{1}{3} & \forall j \in N(i) \\ 0 & \forall j \notin N(i) \end{cases} \tag{2.7}$$

It is also assumed that f() is given by Eqn. 2.1.

Since f() is generally not identically equal to one, there is a finite probability that the algorithm will remain in configuration i. The following equation expresses this probability:

$$P_{ii}(T) = 1 - \sum_{j \in N(i)} P_{ij}(T) \tag{2.8}$$

The stochastic process represented by the evolution of the random variable X produced by the simulated annealing algorithm is a *Markov* process. In fact, Eqns. 2.3, 2.4, and 2.8 imply that the value of X_{k+1} depends only on the value of X_k. That is, if the present state is known exactly, then the probability of any particular future behavior of the process is unaltered by additional knowledge concerning the past behavior. Since the configuration space of combinatorial optimization problems is a countable and, in general, a finite set, the process is a *discrete-time Markov chain with a finite state space.*

The definition of the *radius* of the graph underlying the Markov chain is now presented. Let S_m be the set of all points which are local maxima for the cost function. That is,

$$S_m \equiv \{\, i \in S \mid c(j) \leq c(i) \quad \forall j \in N(i) \,\} \tag{2.9}$$

Then the radius r is defined by:

$$r \equiv \min_{i \in (S - S_m)} \max_{j \in S} d(i, j) \tag{2.10}$$

where $d(i, j)$ is the *distance* of j from i measured by the length, or number of edges, of the shortest path from i to j in G.

A Lipschitz-like constant bounding the local *slope* of the cost function is given by

$$L = \max_{i \in S} \max_{j \in N(i)} |c(j) - c(i)| \tag{2.11}$$

The properties of simulated annealing were studied using Markov chains as the theoretical models. Theoretical analyses have shown that simulated annealing generates with probability one the global optimum of the optimization problem, provided that certain conditions on the number of iterations at each T are satisfied.[1 2] These early analyses used time-homogeneous Markov chains and were required to use unrealistic assumptions on the number of iterations performed at each value of T. Subsequent works based on time-

---

1 F. Romeo and A. Sangiovanni-Vincentelli, "Probabilistic Hill Climbing Algorithms: Properties and Applications," (1985).

2 M. Lundy and A. Mees, "Convergence of the Annealing Algorithm," (1984).

inhomogeneous Markov chains accommodated the time dependence of the temperature.[1 2 3 4 5 6]

Mitra, et.al.[7] proved that for an arbitrary but bounded cost function and an annealing schedule of the form:

$$T_m = \frac{\gamma}{\log(m + m_0 + 1)}, \qquad m = 0, 1, 2, \ldots \tag{2.12}$$

where m_0 is any parameter satisfying $1 \le m_0 < \infty$, the Markov chain is *strongly ergodic* if

$$\gamma \ge rL \tag{2.13}$$

In Eqn. 2.13, r is the radius of the graph underlying the Markov chain and L is a Lipschitz-like constant bounding the slope of the cost function. Strong ergodicity implies that, for any starting probability vector, the state probability vector converges component-wise to a constant vector $\hat{\mathbf{e}}$. Furthermore, it has been shown that $\hat{\mathbf{e}}$ is the *optimum* vector, that is, the vector in which all elements are zero except those with the indices of the least-cost configurations.

The other main result by Mitra, et. al., is on finite-time behavior and the rate of convergence. A bound on the departure of the state vector from the optimum vector after a *finite* number of iterations was derived. This bound indicates how the annealing schedule must be balanced between conflicting requirements for optimum performance. A simple corollary to this result states that for a large number of iterations k, the L_1-norm of the difference of the state vector from the optimum vector is

$$O\left(\frac{1}{k^{\min\{a,\, b\}}}\right)$$

where a and b, respectively, increase and decrease with increasing γ.

A set of results on distributions which are called *quasi-stationary* were also developed by Mitra, et. al.[8] The quasi-stationary distributions are the equilibrium distributions of time-homogeneous Markov chains obtained from simulated annealing by holding the temperature fixed at various values. The dependence of the quasi-stationary distributions on temperature was shown to have a number of desirable properties. These properties are essential for the analysis of the time-inhomogeneous Markov chains obtained from annealing

---

[1] D. Geman and S. Geman, "Stochastic Relaxation, Gibbs Distributions, and the Bayesian Restoration of Images," (1984).

[2] B. Gidas, "Non-Stationary Markov Chains and Convergence of the Annealing Algorithm," (1985).

[3] B. Hajek, "Cooling Schedules for Optimal Annealing," (1985).

[4] D. Isaacson and R. Madsen, "Markov Chains: Theory and Applications," (1976).

[5] M. Iosifescu, "Finite Markov Processes and their Applications," (1980).

[6] E. Senata, "Non-Negative Matrices and Markov Chains," (1980).

[7] D. Mitra, F. Romeo, and A. Sangiovanni-Vincentelli, "Convergence and Finite-Time Behavior of Simulated Annealing," (1985).

[8] D. Mitra, F. Romeo, and A. Sangiovanni-Vincentelli, "Convergence and Finite-Time Behavior of Simulated Annealing," (1985).

schedules given by Eqn. 2.12. Furthermore, they are of independent interest since they hold for annealing schedules considerably more general than that given by Eqn. 2.12. This may be important if it is possible to design annealing schedules matched to special properties of the cost function.

2.4 Overview of Work on General Annealing Schedules

The previous section presented proof that, under certain conditions, simulated annealing will converge to the global optimum with probability one. Unfortunately, these results were asymptotic in the sense that an infinite amount of computation time would be necessary to guarantee the convergence to the global optimum. Consequently, there has been much research on practical implementations of simulated annealing for solving general combinatorial optimization problems.[1] [2] [3] [4] [5] In order to obtain the best results, we often find that the cost function and the generation of new configurations function are problem specific. Consequently, the term *general* as applied to the simulated annealing schedule implies that the initial temperature, the cooling schedule, and the convergence criterion are controlled in a problem independent manner. The cooling schedule consists of a sequence of monotonically decreasing temperatures and the condition of equilibrium at each temperature. There have been a few problem-independent, general annealing methods reported recently.[6] [7] [8] [9] In the remainder of this subsection, an overview of the method by Huang, et. al. is presented.[10]

The conditions determining an annealing schedule are: (1) the initial temperature, (2) the temperature decrement, (3) the equilibrium condition, and (4) the stopping, or convergence criterion. For an annealing schedule to be problem independent, the parameters used in the four conditions should be determined by the algorithm itself and should not have any predefined values. For an efficient cooling, an early detection of equilibrium and

---

[1] M. Huang, F. Romeo, and A. Sangiovanni-Vincentelli, "An Efficient General Cooling Schedule for Simulated Annealing," (1986).

[2] D. Geman and S. Geman, "Stochastic Relaxation, Gibbs Distributions, and the Bayesian Restoration of Images," (1984).

[3] J. Lam and J. M. Delosme, "Logic Minimization Using Simulated Annealing," (1986).

[4] E. Aarts and P. Laarhoven, "Statistical Cooling: A General Approach to Combinatorial Optimization Problems," (1985).

[5] R. Otten and L. Ginneken, "Floorplan Desing Using Simulated Annealing," (1984).

[6] R. Otten and L. Ginneken, "Floorplan Desing Using Simulated Annealing," (1984).

[7] E. Aarts and P. Laarhoven, "Statistical Cooling: A General Approach to Combinatorial Optimization Problems," (1985).

[8] M. Huang, F. Romeo, and A. Sangiovanni-Vincentelli, "An Efficient General Cooling Schedule for Simulated Annealing," (1986).

[9] J. Lam and J. M. Delosme, "Logic Minimization Using Simulated Annealing," (1986).

[10] M. Huang, F. Romeo, and A. Sangiovanni-Vincentelli, "An Efficient General Cooling Schedule for Simulated Annealing," (1986).

an aggressive temperature decrement are desirable. For a reliable cooling, the temperature should not be decremented until equilibrium is reestablished.

2.4.1 The initial temperature

The condition proposed by White is used to determine the starting temperature.[1] Letting σ represent the standard deviation of the cost distribution, White argued that an appropriate initial temperature T_0 is such that

$$\sigma \ll T_0$$

To determine σ, an initial exploration of the configuration space is performed. During the exploration, all generated configurations are accepted. The standard deviation of the cost distribution is computed and the initial temperature is set according to

$$T_0 = k\sigma$$

The value of k is determined by assuming a normal distribution and selecting a temperature high enough to accept with probability P a configuration whose cost is 3σ worse than the present configuration. This leads to the following expression for k :

$$k = -\frac{3}{\ln P}$$

2.4.2 The temperature decrement

Approaches have been proposed which use σ in the determination of the next temperature decrement.[2 3] The advantage of these approaches is that the temperature is dynamically controlled by the annealing process itself, and is therefore applicable to a wide variety of problems. Although a notion of quasi-equilibrium was used to guide the temperature decrement, the amount of the decrement was not clearly specified, nor were there any reliable guidelines specifying the control of the decrement.

The approach by Huang, et. al. used the so-called *annealing curve*, a plot of the average cost $\bar{C}$ versus the log of the temperature, to guide the temperature decrease.[4] Their idea was to control the temperature such that $\bar{C}$ decreases in a uniform manner. The slope of the annealing curve is given by:

$$\frac{d\bar{C}}{d\ln(T)} = T\frac{d\bar{C}}{dT}$$

Using the following relationship,[5]

---

[1] S. White, "Concepts of Scale in Simulated Annealing," (1984).

[2] E. Aarts and P. Laarhoven, "Statistical Cooling: A General Approach to Combinatorial Optimization Problems," (1985).

[3] R. Otten and L. Ginneken, "Floorplan Desing Using Simulated Annealing," (1984).

[4] M. Huang, F. Romeo, and A. Sangiovanni-Vincentelli, "An Efficient General Cooling Schedule for Simulated Annealing," (1986).

[5] F. Reif, "Statistical and Thermal Physics," (1965).

$$\frac{d\bar{C}}{dT} = \frac{\sigma^2}{T^2}$$

it follows that

$$\frac{d\bar{C}}{d\ln(T)} = \frac{\sigma^2}{T}$$

Using a linear approximation of the slope,

$$\frac{\Delta C}{\ln(T') - \ln(T)} = \frac{\sigma^2}{T}$$

where ΔC is the difference in cost at temperatures T and T'. Rearrangement of the previous equation results in:

$$T' = T \exp\left(\frac{T\,\Delta C}{\sigma^2}\right)$$

Huang, et. al., argued that quasi-equilibrium could be maintained by requiring the expected decrease in average cost to be less than σ. That is, let $\Delta C = -\lambda\sigma$, where $\lambda \leq 1$. Thus,

$$T' = T \exp\left(-\frac{\lambda T}{\sigma}\right)$$

2.4.3 The equilibrium condition

Equilibrium in this context implies the establishment of a steady-state probability distribution of the accessible states. Huang, et. al., base their equilibrium condition on the following observation: When equilibrium is established, the ratio of the number of new configurations generated whose cost is within a range δ from the average cost to the total number of newly accepted configurations will reach a stable value χ. This value depends on the nature of the cost distribution and the sampling range δ. At high temperatures, the cost distribution is close to a normal distribution. In this case, the ratio of the number of configurations whose cost is within the range $(\bar{C} - \delta, \bar{C} + \delta)$, referred to as the *within count*, to the number of accepted configurations is given by

$$\mathrm{erf}\left(\frac{\delta}{\sigma}\right)$$

where erf() is the error function.[1] Based on this value and the problem size, a target *within count* and a maximum tolerance limit are established as the equilibrium parameters. Equilibrium is considered to be maintained if the *within count* reaches its target value before the number of acceptances outside the designated range reaches the maximum tolerance limit. If the maximum tolerance count is exceeded before the *within count* is attained, both counts are reset to zero,and the counting is resumed.

The parameter δ is set to be a fraction of σ so that the final configuration at any temperature is a configuration whose cost is close to the average cost (within the range of δ) and therefore is a highly accessible state at that temperature. In Huang, et. al., a typical

[1] W. Feller, "An Introduction to Probability Theory and Applications," (1970).

value for δ is 0.5σ, in which case $\chi = \text{erf}(0.5) = 0.38$. The *within count* was then set to: 0.38 (3 *N*), where *N* is the size of problem. The maximum tolerance was set to: 0.62 (3*N*).

2.4.4 The stopping, or convergence, criterion

When equilibrium is established, the maximum and minimum costs among the accepted states at that temperature are compared with the maximum change in cost in any accepted move. If they are the same, apparently all the accessible states have comparable costs and there is no need to continue the simulated annealing process. The temperature is then set to zero and the algorithm becomes a standard *greedy* random selection algorithm.

2.5 Implementations of Simulated Annealing for Placement and Global Routing

In this section, I will overview the general implementation details for simulated annealing as applied to placement and global routing of integrated circuits. In this work, I have focused attention on the implementation of the cost and generate functions. These functions have a very strong influence on the quality of the results yielded by an implementation of simulated annealing. For example, in the case of simulated annealing applied to placement problems, it appears that the implementation of these two functions sets a lower bound on the best possible solution that can be obtained for the final total interconnect length. That is, small changes in the other simulated annealing functions tend to yield virtually unnoticeable changes in the results. On the other hand, small changes to the cost function, or to the generation of states function, can yield substantial changes in the results. In other words, configuration-space topology changes have the largest impact on the performance of an implementation of simulated annealing.

The general implementation details of the five basic simulated annealing functions which are shared by several of the placement and global routing programs are now briefly reviewed.

1. Generating New States: The implementations begin with a random initial placement or wiring configuration. A new state is generated by either exchanging two fundamental units or by moving a unit to another location. For the gate array placement program presented in the Appendix, the new state is generated by the interchange of two *modules*, where a module refers to a fundamental unit specified in the net list. The standard cell placement program also generates new states by the interchange of cells. However, because standard cells typically vary in width, the interchange of two cells often results in a non-feasible solution because overlaps are obviously not allowed. This is solved by a penalty function approach, first described by Kirkpatrick, Gelatt, and Vecchi.[1] The implementation of this approach will be described in Chapters 3 and 4. The penalty function approach was also employed by the macro/custom cell placement program because the cells typically vary in both height and width.

For the standard cell and macro/custom cell problems, new states are also generated by the movement of a cell to a new location. The use of both methods of generating new

[1] S. Kirkpatrick, C. Gelatt, and M. Vecchi, "Optimization by Simulated Annealing," (1983).

states is necessary to achieve the best results. Furthermore, orientation changes of standard and macro/custom cells are performed which result in new states. New states can also be generated for custom cells by assigning a new location to a pin or group of pins, and by changing the aspect ratio of the cell. For the standard cell global router, new states by assigning a portion of a net to a different channel.

2. *Cost Function*: The cost function for the placement programs is based on the total interconnect length. The standard cell and macro/custom cell algorithms also include a penalty function term which penalizes overlaps of the cells. The cost function for the standard cell global router is based on the estimated wiring area which is approximated by the total channel density, that is, the sum over all channels of the channel density.

3. *Generating New Values of T*: In the experiments I conducted, the best results were obtained when α is the largest, approximately 0.95, during the stages of the algorithm when the value of the cost function is decreasing rapidly. Furthermore, the value of α is given its lowest value, usually 0.80, during the initial and latter stages of the annealing process. The value of α is gradually increased from its lowest value to its highest value, and then gradually decreased back to its lowest value.

4. *The Inner Loop Criterion*: The inner loop criterion was implemented by the specification of the number of new states generated for each stage of the annealing process. This number is specified as a multiple of the number of fundamental units for the placement or global routing problem. Typically, for current problem sizes, on the order of 100 new states are generated per unit at each stage.

5. *The Stopping Criterion*: The stopping criterion was implemented by recording the value of the cost function at the end of each stage of the annealing process. It is satisfied when the value of the cost function has remained unchanged at the end of three consecutive stages.

2.6 The Function f()

Kirkpatrick, Gelatt, and Vecchi proposed the following expression for f(), which resembles a Boltzmann factor:[1]

$$\mathrm{f}(\Delta c, T) = \min\left\{ 1.0,\ \exp\left(-\frac{\Delta c}{T}\right)\right\} \tag{2.14}$$

I also tested another form for f(), which resembled the Fermi-Dirac factor:

$$\mathrm{f}(\Delta c, T) = \frac{1}{1 + \exp\left(\frac{\Delta c}{T}\right)} \tag{2.15}$$

[1] S. Kirkpatrick, C. Gelatt, and M. Vecchi, "Optimization by Simulated Annealing," (1983).

It has been shown that this functional form for f() is sufficient to prove the convergence of simulated annealing.[1] Note that for Eqn. 2.15, new states which *weakly* reduce the value of the cost function may be rejected. For Eqn. 2.14, new states which reduce c are always accepted. Intuitively, rejecting new states which only weakly reduce the cost function appears promising. That is, it would appear that the algorithm would focus on accepting only those moves which significantly reduce the cost.

Unfortunately, this intuition appears not to be valid. For example, Table 2.1 has data obtained from the row-based placement program described in Chapter 3 in which the initial total interconnect length was 1,256,154 for a 1500-cell circuit. The difference in the results produced by the two approaches is not great. Eqn. 2.14 resulted in an average improvement of 1.1% over Eqn. 2.15.

No. Trials	f()	Avg. Final Interconnect Length
2	Eqn. 2.14	682,084
2	Eqn. 2.15	689,498

Table 2.1 Comparison of the two forms for **f()**.

Additional comparative tests between the two forms for **f()** were performed on the *island-style* gate array placement program presented in the Appendix. The tests were completed on a very small gate array in which it was known that the global optimum solution had a cost value of 38. The temperature was reduced by a constant factor $\alpha = 0.95$.

No. Trials	f()	Avg. Final Cost Value
40	Eqn. 2.14	39.60
40	Eqn. 2.15	40.63

Table 2.2 Second comparison of the two forms for **f()**.

Note that the Boltzmann-like factor for **f()** produced a better average final cost, a slightly more favorable worst solution, and a greater number of optimal final solutions.

A second set of tests was also performed in which $\alpha = 0.99$ was used.

[1] D. Mitra, F. Romeo, and A. Sangiovanni-Vincentelli, "Convergence and Finite-Time Behavior of Simulated Annealing," (1985).

No. Trials	f()	Avg. Final Cost Value
40	Eqn. 2.14	38.75
40	Eqn. 2.15	38.85

Table 2.3 Third comparison of the two forms for f().

Again, the Boltzmann-like factor for f() produced a better average final cost and a greater number of optimal final solutions. In summary, Eqn. 2.15 appears to perform *almost* as well as Eqn. 2.14, but never better. Since Eqn. 2.14 has a simpler form, this function was used for f().

2.7 Fast Evaluation of the Exponential Function

The basic simulated annealing algorithm requires an evaluation of

$$\exp\left(-\frac{\Delta c}{T}\right)$$

each time a new state is generated. Each evaluation of the exponential function requires 107us of CPU time on a VAX-11/780 system running VMS and 75us on an IBM-3081K system running UTS. For 10^8 evaluations of the exponential function, the total CPU time would be 2.78 hours for the VAX-11/780 and 2.08 hours for the IBM 3081K.

Substantial reductions in the computation time for the evaluation of the exponential function were achieved by using a table look-up technique.[1] This technique requires only 3 table look-ups, 2 floating multiplies , and 2 shifting operations to achieve good accuracy. It will be indicated below that the result can be different from the exact value by at most a factor of 1.000000119. The computation time for the evaluation of the exponential function was reduced to 44us on the VAX-11/780 and to 2.5us on the IBM 3081K as a result of this table look-up technique.

The details of the technique are now presented. First, note that the function to be evaluated is always

$$\exp\left(-\frac{\Delta c}{T}\right)$$

where both Δc and T are positive quantities since new states characterized by $\Delta c \leq 0$ are always accepted. Assume that it is desired to compute e^x where

$$x = -\frac{\Delta c}{T}$$

Although $x < 0$ is the relevant case in simulated annealing, $x > 0$ will be treated here for simplicity. This is not a limitation since

$$e^{-x} = \frac{1}{e^x}$$

[1] J. Deutsch, Private Communication, (1984).

The largest floating point number that can be stored in the D_Floating format on the VAX-11/780 is 1.7×10^{38}. This implies that a value of x that exceeds 88 will be essentially ∞. Therefore, the technique first checks to see if its argument exceeds 88. If so, it returns ∞. Furthermore, in an effort to further reduce the computation time, if x is less than 10^{-7}, e^x is approximated as $(1 + x)$.

It follows that the range of x for which there is interest in evaluating e^x is then: $10^{-7} < x < 88$. Rewritten, the range of interest is: $2^{-23} < x < (2^7 - 40)$. In other words, the seven bits to the left of the decimal point in the base two floating point representation of x and the 23 bits to the right of the decimal point. That is, a total of 30 bits are of interest. If x is multiplied by 2^{23} (this can be thought of as a left shift by 23 bit positions), then the 30 bits are all to the left of the decimal point. In this case, the floating point number can be directly converted and stored as a 32-bit integer such that the 30 least significant bits are of interest.

If substantially more than 3072 evaluations of the exponential function will be required during the course of a run, then the following procedure reduces the evaluation time. Evaluate e^x by using

$$e^x = e^{x_1 + x_2 + x_3} = e^{x_1} e^{x_2} e^{x_3}$$

where x is written as

$$x = x_1 + x_2 + x_3$$

At the start of the program, three sets of 1024 evaluations using the exact exponential function are performed. The three sets are broken up as follows, where x represents the 30 bits of interest:

$$\begin{aligned} \text{set } 1 &= \{ e^x \mid 2^{-3} < x \le 2^7 \} \\ \text{set } 2 &= \{ e^x \mid 2^{-13} < x \le 2^{-3} \} \\ \text{set } 3 &= \{ e^x \mid 2^{-23} < x \le 2^{-13} \} \end{aligned} \tag{2.16}$$

The first set is stored in **table1**, the second set is stored in **table2,** and the third set is stored in **table3**. There are three steps in the evaluation procedure. (1) The most significant group of 10 bits of x (bits 29 through 20) are shifted right by 20 bits and the ten bits are masked out. This quantity determines *index*1, the index into **table1**. Each entry in the table has the value of

$$\exp\left(index1 \cdot 2^{-3} \right)$$

(2) The next most significant group of 10 bits of x (bits 19 through 10) are shifted right by 10 bits and the ten bits are masked out. This quantity determines *index*2, the index into **table2**. Each entry in the table has the value of

$$\exp\left(index2 \cdot 2^{-13} \right)$$

(3) The least significant group of 10 bits of x (bits 9 through 0) are masked out. This quantity determines *index*3, the index into **table3**. Each entry in the table has the value of

$$\exp\left(index3 \cdot 2^{-23} \right)$$

The value of e^x is then given by:

$$e^x = \textbf{table1}[\ index1\ ] \times \textbf{table2}[\ index2\ ] \times \textbf{table3}[\ index3\ ] \tag{2.17}$$

This technique requires only 3 table look-ups, 2 floating multiplies , and 2 shifting operations. With respect to the accuracy of the method, note that in the original floating point representation of x, only the least significant bits representing the powers of 2 descending from 2^{-24}, 2^{-25},2^{-26}, $\cdots$, are omitted. Note that the exact value of x can differ from the approximated value by at most $(2^{-23} - 1)$. The exponential of this value is approximately $(1 + 1.192093 \times 10^{-7})$. Hence the conclusion that the result can be different from the exact value by at most a factor of 1.000000119.

Chapter 3

Placement and Global Routing of Standard Cell Integrated Circuits

3.1 Introduction

This chapter presents the algorithms and the implementation of a standard cell placement and global routing package. This package, named *TimberWolfSC*,[1] makes extensive use of the simulated annealing algorithm presented in the previous chapter. The generalized standard cell layout style handled by TimberWolfSC is characterized by horizontal rows of standard cells with pads placed around the periphery of the chip. Furthermore, macro blocks may be present on the chip. An example of such a standard cell layout is shown in Figure 3.1.

Note that the standard cells are permitted to have varying heights. In Figure 3.1, the cells in rows 1, 2 and 3 (numbered starting from the bottom), as well as in portions of rows 6 and 8 have greater height than the other cells. Macro blocks are also allowed on the same chip. In particular, note the macro block named *Cell MBB* which lies below the rows. Also, macro blocks named *Cell MUL* (upper left corner) and *Cell MM* (middle of the layout) are illustrated in Figure 3.1. The placement of eighteen pads along the periphery of the chip is also shown in Figure 3.1.

[1] All references to TimberWolfSC in this chapter and in this book refer to version 3.2. Work on TimberWolfSC has continued; for example, see: C. Sechen and K. W. Lee, "An Improved Simulated Annealing Algorithm for Row-Based Placement," (1987).

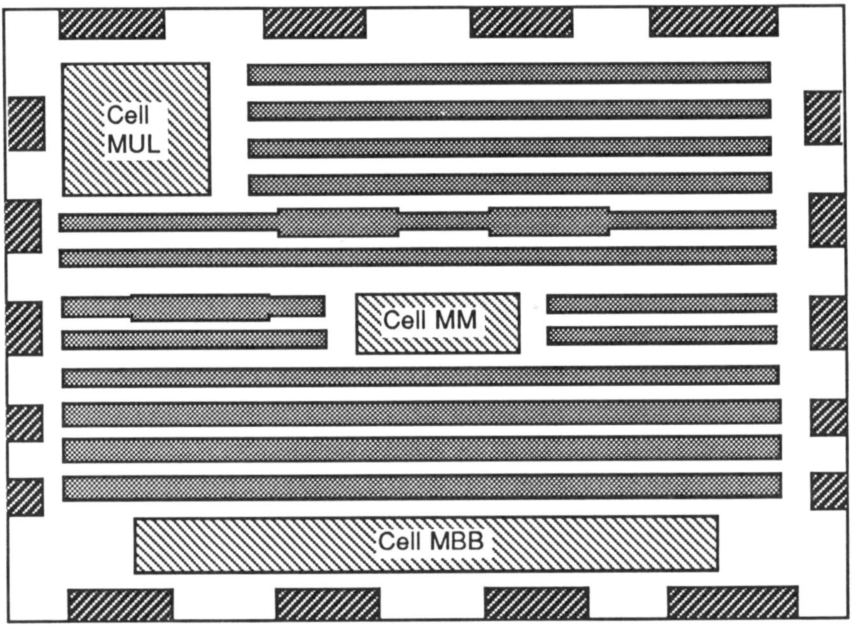

Figure 3.1 The generalized standard cell layout style handled by TimberWolfSC.

The main objectives in standard cell layout are to maximize the performance of the chip (for example, keeping critical nets as short as possible) and then to minimize the area of the chip. Most of the placement algorithms developed for standard cell placement use as the objective function the sum of the estimated net lengths. Following the placement, the circuit must then be globally routed in which the objective is to minimize the height of the horizontal channels (between the rows).

Previous placement methodologies were either: (1) Manual placement, which has generally resulted in area and performance efficiency for small circuits. However, for very large circuits, not only is the design time prohibitively long, but the area and performance suffer. (2) Automatic placement, which has generally resulted in relatively poor area efficiency (in comparison to manual placement) and furthermore, the algorithms used have usually not had the amount of flexibility and extensibility desired by the users. With respect to global routing, relatively few algorithms have been previously reported for the standard cell layout style.

During the course of the development of the TimberWolfSC package, the overriding objective has been to develop a complete placement and routing package with quality algorithms and with proven results. The main focus has been to achieve the best possible results while possibly sacrificing computation time. Another main objective was to develop a flexible and extensible placement and routing package which is applicable to state-of-the-art industrial circuits. As a consequence of these goals, extensions and modifications of the general combinatorial optimization technique known as simulated annealing were selected as the primary algorithmic techniques.[1] Several of the algorithms in TimberWolfSC are based on simulated annealing.

This chapter presents the algorithms and their implementation, as well as the results for TimberWolfSC, a standard cell placement and global routing package. The earliest version of TimberWolfSC, version 1.0, consisted only of the first stage of version 3.2.[2] [3] The standard cell global router is now an integral part of the general placement procedure. Additional flexibility was also added to TimberWolfSC. The placement and global routing now proceed over 3 distinct stages. In the first stage, TimberWolfSC places the cells while seeking to minimize the total estimated interconnect cost. This stage of the placement is performed using simulated annealing. In the second stage of the placement, TimberWolfSC inserts feed-through cells as required and the minimization of the total estimated interconnect cost proceeds again in the manner of simulated annealing. Furthermore, TimberWolfSC utilizes uncommitted feed throughs built in to the standard cells whenever possible to avoid the addition of a feed-through cell. The second stage comes to a close following a global routing step, in which the number of wiring tracks needed is accurately estimated. During the third and final stage, local changes are made to the placement should such a change lead to a reduction in the number of wiring tracks required.

Results on industrial circuits versus numerous automatic and manual layout methods showed that TimberWolfSC yielded area savings ranging from 15 to 75 percent. For all circuits tested, the global router portion of stage 2 reduced the number of wiring tracks needed by an additional 7 to 16 percent in comparison to stage 1 alone. Furthermore, the combination of the global router of stage 2 and stage 3 resulted in wiring track reductions ranging from 15 to 25 percent in comparison to stage 1 alone. TimberWolfSC also features critical-net weighting for performance-driven placement.

The remainder of this chapter is organized as follows: Section 3.2, presents some details of the general TimberWolfSC methodology. In Section 3.3, the algorithm for the first stage of the standard cell placement and global routing program is described. Section 3.4 presents the algorithmic details for stage 2, including the standard cell global router. Next, Section 3.5 presents the placement-refinement algorithm of stage 3. Finally, a summary of the experimental results is presented in Section 3.6.

3.2 The General TimberWolfSC Methodology

TimberWolfSC is applicable to the generalized standard cell placement and global routing problem. An example of a generalized standard cell placement configuration is shown in Figure 3.1. TimberWolfSC constructs the standard cell circuit configuration based on the net list data.

It is usually the case that the longest row determines the width of the standard cell chip. It is very important that the sum of the lengths of the rows is equal to the sum of the widths of the standard cells. If this is not the case, the placement algorithm is likely to use

---

[1] S. Kirkpatrick, C. Gelatt, and M. Vecchi, "Optimization by Simulated Annealing," (1983).

[2] C. Sechen and A. Sangiovanni-Vincentelli, "The TimberWolf Placement and Routing Package," (1984).

[3] C. Sechen and A. Sangiovanni-Vincentelli, "The TimberWolf Placement and Routing Package," (1985).

to excess a particular row, causing it to be the longest. This is particularly true for iterative placement algorithms, such as the simulated annealing algorithm used in TimberWolfSC.

Furthermore, for some technologies, usually single-metal although not exclusively so, in order to route through a row it is necessary to insert a *feed-through cell* into the row at the desired cross-over point. Statistically, it is not surprising that a majority of the added feed-through cells must be placed into the middlemost rows of the standard cell layout. The feed-through cells are usually not very wide; however, in great numbers, they can contribute heavily to the length of a row. In order to avoid having the middle rows bulge out excessively, for some circuits and technologies it is best to anticipate the large demand for feed-through cells and allow somewhat fewer standard cells to be placed into the middle rows.

Experience gained from industry indicated that, in almost every case, the placement of pads and macro blocks was predetermined by the circuit designers. Consequently, in TimberWolfSC, pads and macro blocks are placed in accordance with net list specifications, and are not changed by the simulated annealing algorithm.

The above considerations make it imperative that the exact placement of the rows (with their respective assigned lengths), macro blocks and pads be done by TimberWolfSC. That is, the net list data is used only to specify *general* localizations as to desired placement. For example, a macro block may be specified to lie in the *upper left corner* of the chip, or a group of pads may be given a specified *order* and may be designated to lie on the *top* side of the chip. In this fashion, TimberWolfSC can ensure that the initial configuration of the chip is such that the simulated annealing algorithm will be best able to find the placement of the highest possible quality every time.

3.2.1 Finding the optimal target row lengths

As presented above, the proper selection of the target lengths of the rows is of utmost importance to assure optimum performance by the simulated annealing algorithm. In this subsection, the algorithm for computing the target row lengths is given. Some of the parameters extracted from the net list include: (1) the number of rows, expressed by N_r, (2) the center-to-center vertical row-separation distance, expressed by d_s, (3) the height of the rows, h_r (taken as an average of the actual row heights), and (4) the number of standard cells, N_c. Also extracted from the net list is a parameter, one for each row $r \in \{1, \cdots, N_r\}$, specifying a desired *relative row length*, $l_r(r)$, for each r. As an example, note in Figure 3.2 that the upper six rows are specified to have a longer length than the bottom four rows.

Figure 3.2 The upper six rows are specified to have a longer relative row length than the bottom four rows.

The set of $l_r(r)$ need not be normalized, and in fact, it is common to characterize the default row length as having $l_r(r) = 1.0$. The length of row per unit of relative row length may be expressed as:

$$L_r = \frac{\sum_{c=1}^{N_c} w_c[c]}{\sum_{r=1}^{N_r} l_r(r)} \tag{3.1}$$

where $w_c[c]$ is the width of cell c.

Yet another parameter is included in the net list to reflect the expected number of feed-through-cell instances which will need to be added to the rows. As mentioned above, it is expected that more feed-through cells will be placed in the middle rows. This effect is modeled by the specification, in the net list, of an indentation I_m. For the case of equal target row lengths, I_m represents the ratio of the length of the middle row to the length of the top (or bottom) row. The lengths of the rows between the middle row m and the top row t, and between m and the bottom row b are assumed to be linearly distributed. This resulted from the assumption that the distribution of feed-through cells follows this linear pattern. A proof of this assumption was not attempted, although it appeared to be valid for all of the industrial circuits which were tested. Note that $0 < I_m \leq 1$. Then the ratio of the length of row r at y-coordinate y_r to the length of the top row is given by the indentation factor $i_r(y_r)$:

$$i_r(y_r) = I_m + \frac{(1 - I_m)}{(y_t - y_m)} \left| y_r - y_m \right| \tag{3.2}$$

where y_t is the y-coordinate of the top row. Note that y_m is precisely the y-coordinate of the middle row if N_r is odd and if the separation between all rows is d_s; otherwise,

$$y_m = \frac{y_b + y_t}{2}$$

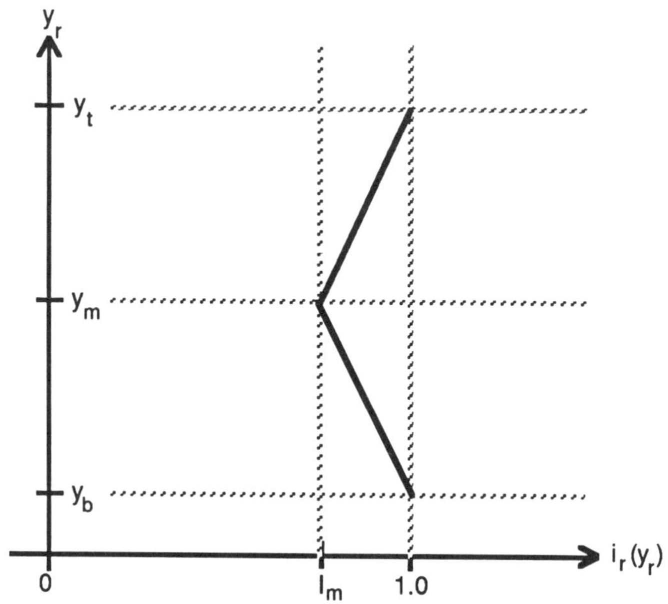

Figure 3.3 The piece-wise linear row-length modulation function $i_r(y_r)$.

The originally intended length of a row r is given by $L_r{\cdot}l_r(r)$. The actual desired length of row r is modulated by the influence of the required indentation factor. That is, some rows have their lengths increased beyond $L_r{\cdot}l_r(r)$ and others have their lengths reduced. The desired length of row r, expressed by $l(r)$, can be expressed as:

$$l(r) = L_r{\cdot}l_r(r)\,(1 + \Delta_r) \tag{3.3}$$

where Δ_r reflects the fact that a row r will have its desired length adjusted due to the indentation factor. The value of Δ_r is given by:

$$\Delta_r = i_r(y_r) - I_m - i_0$$

Note that $i_r(y_r) \geq I_m$ for all r, as indicated in Figure 3.3. Hence, the constant i_0 reflects the fact that some rows must have their lengths increased and others have their lengths reduced. Conservation of total row length determines the value of i_0 in the following manner:

$$\sum_{r=1}^{N_r} L_r{\cdot}l_r(r) = \sum_{r=1}^{N_r} l(r) = \sum_{r=1}^{N_r} \{\, L_r{\cdot}l_r(r)\,[1 + \Delta_r]\,\}$$

Substitution for Δ_r then yields:

$$\sum_{r=1}^{N_r} l_r(r) = \sum_{r=1}^{N_r} \left\{ l_r(r)\left[1 + \frac{(1 - I_m)}{(y_t - y_m)}\left(|y_r - y_m| - i_0 \right) \right] \right\} \tag{3.4}$$

This equation is then solved for i_0 to yield:

$$i_0 = \frac{\sum_{r=1}^{N_r} l_r(r) \frac{(1 - I_m)}{(y_t - y_m)} |y_r - y_m|}{\sum_{r=1}^{N_r} l_r(r)} \tag{3.5}$$

The basic algorithmic strategy is to first compute i_0 by means of Eqn. 3.5. Then substitute i_0 into the following equation to yield the desired length of each row r:

$$l(r) = L_r \cdot l_r(r) \left\{ 1 + \frac{(1 - I_m)}{(y_t - y_m)} |y_r - y_m| - i_0 \right\} \tag{3.6}$$

In Figure 3.2, the bottom four rows have a relative length which is less than that of the top six rows. Furthermore, in this figure, I_m is set to 1.0. For $I_m < 1.0$, the example of Figure 3.2 assumes a row configuration of the form shown in Figure 3.4. The vertical placement of the rows is determined by the net list values of d_s and h_r.

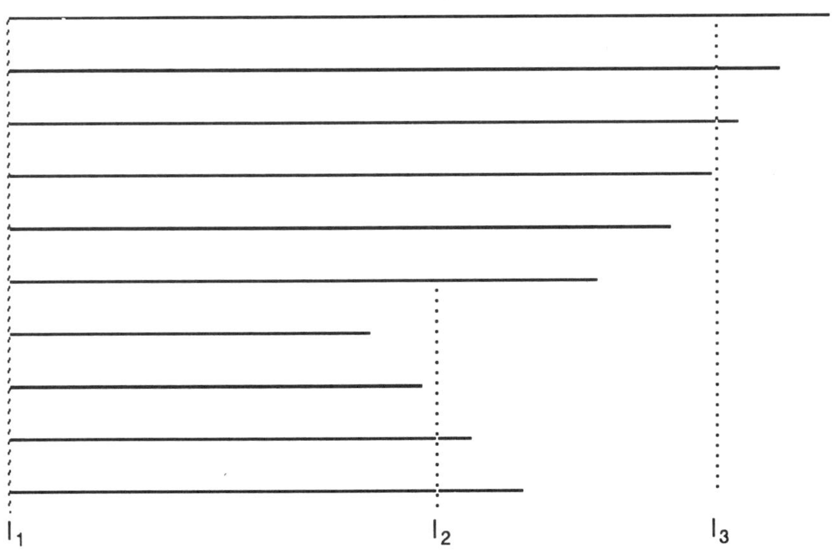

Figure 3.4 The row configuration obtained by setting $I_m < 1.0$ for the example of Figure 3.2.

In my experience of placing industrial circuits, it is usually the case that a target aspect ratio is desired for the core area. However, the separation between the rows can only be estimated with a complete global routing step, which cannot be completed realistically until the placement is done. Hence the usual approach requires some iteration between selecting the number of rows and monitoring the aspect ratio which results from the actual placement and global routing. It would be helpful if the placement program could select automatically the number of rows which would result in the desired aspect ratio. However, this could only be done if global routing was performed during placement, which would impact drastically the run time.

3.2.2 Critical-net weighting

TimberWolfSC features critical-net weighting and the ability to weight the *horizontal span* separately from the *vertical span* for each net. The horizontal span of a net is defined as the horizontal span of the bounding box of the pins comprising the net. Similarly, the vertical span of a net is defined as the vertical span of the bounding box of the pins comprising the net. The Total Estimated Interconnect Length (TEIL) is the sum over all nets of the horizontal span plus the vertical span of the net. The Total Estimated Interconnect Cost (TEIC) is given by the sum over all nets of: (1) the horizontal span of the net times the horizontal weighting factor for the net, and (2) the vertical span of the net times the vertical weighting factor for the net.

3.3 The Algorithm for Stage 1 of TimberWolfSC

This section presents the algorithmic details for the first stage of TimberWolfSC. The purpose of stage 1 is to find a placement of the standard cells while seeking to minimize the TEIC. A simulated annealing algorithm is employed.

3.3.1 The cost function

The cost function for the simulated annealing algorithm of stage 1 consists of three independent terms. The stage 1 algorithm seeks to minimize the first term while driving the next two terms of the cost function toward zero.

3.3.1.1 The first term in the cost function

The first term is a function $C_1()$ returning the TEIC. The following terminology is used: (1) N_n is the total number of nets, (2) $X_s(n)$ returns the span of net n in the x, or horizontal direction, (3) $Y_s(n)$ returns the span of net n in the y, or vertical direction, (4) $H_w(n)$ returns the weighting factor for net n for the horizontal direction, and (5) $V_w(n)$ returns the weighting factor for net n for the vertical direction. The function $C_1()$ is then given by:

$$C_1() = \sum_{n=1}^{N_n} \{ X_s(n) \cdot H_w(n) + Y_s(n) \cdot V_w(n) \} \tag{3.7}$$

3.3.1.2 The second term in the cost function

The second term is an overlap penalty function. This penalty function was incorporated because of the usual difference in width of the standard cells. In the next section it will be seen that quite often two cells are selected for interchange which differ in width. An exchange of location of these two cells often results in some overlap with one or more of the other cells. Furthermore, a single cell is often selected for a displacement to a random new location. Since there is no prior check for the existence of one or more cells in the vicinity of this new location, some overlapping may well result.

The alternative to the overlap penalty function approach is, of course, to not permit overlapping. For example, when inserting a cell into a row, if insufficient space is available, then the cells to the right are all shifted farther to the right as necessary. This has the obvious disadvantage of destroying the optimization process between the shifted cells and the cells on the neighboring rows. The overlap penalty function approach was employed so as to prevent the disturbance of the placement of the remaining cells when performing an interchange of cells or a displacement of a single cell.

The second term is $C_2()$, and the value it returns is given by:

$$C_2() \; = \; p_2 \sum_{i \neq j} \{ \, O_l(i,j) \, \}^2 \tag{3.8}$$

The function $O_l(i,j)$ returns the amount of overlap of cells i and j in the x direction. If two cells overlap, a penalty is assigned which is proportional to the square of the amount of overlap in the horizontal direction. The amount of overlap is only computed for one dimension since it is often the case that standard cells have the same height.

The normalization parameter p_2 is required since function $C_1()$ scales linearly with the grid size of the net list data, while function $C_2()$ is quadratic with respect to the grid size. For a one micron grid size, $p_2 = 1$ is sufficient to drive the penalty function $C_2()$ toward zero at the conclusion of the first stage.

The implementation of $C_2()$ requires very many queries of the form: Given a cell $j \in \{1, \cdots, N_c\}$ find the set $\{J\}$ of other cells such that:

$$\{J\} \; = \; \{ \, j \in \{1, \cdots, N_c\} \; | \; O_l(i,j) \neq 0 \, \}$$

The search for cells overlapping with a given cell is limited to those other cells on the same row as cell i. However, it is quite possible for more than 500 cells to appear on a given row. Furthermore, it is also quite possible (as will be inferred later) that on the order of 10^7 queries need to be answered regarding the search for cells overlapping with a given cell. It is apparent that a very substantial amount of execution time can be invested in performing overlap calculations. As a result, the time complexity of performing overlap calculations is a crucial issue. A reasonable assumption to make is that aspect ratios of standard cell chips are usually near 1.0. It is then reasonable to conclude that the number of cells in a given row is of order $O\left(\sqrt{N_c}\right)$. The time complexity of determining which cells overlap with a cell i is then $O\left(\sqrt{N_c}\right)$ if all cells in the row of i are checked.

Keeping a sorted list of cells for each row is not an attractive alternative because of the fact that cell overlapping is permitted. It would be very difficult to generate a precise ordering which would permit the determination of the set $\{J\}$ in constant time. Sorting the cells by their centers, given that the cell widths vary substantially, would not appear to be of much help since a cell whose center is well outside the range of interest could still overlap this range.

In TimberWolfSC, the time complexity of overlap calculations was reduced by employing a *bin* storage technique. In this method, the span of each row is divided into a union of N_B non-intersecting bins of equal horizontal span. This implies that every x

coordinate in row r belongs to, or *hashes* to, exactly one of the bins in row r. Given a cell i in row r, the bin assignment for cell i is determined in the following manner: (1) Hash the left edge of cell i to obtain bin b_l^r. (2) Hash the right edge of cell i to obtain bin b_r^r. (3) If $b_l^r = b_r^r$, then b_l^r constitutes the bin assignment for cell i in row r. (4) If $b_l^r \neq b_r^r$, then cell i is not contained by a *single* bin. That is, one or more bin boundaries are overlapped by cell i. In this case, cell i is assigned to bin 0, reflecting the fact that i is not contained by any of the N_B bins. Bin 0 is usually referred to as the *bucket.*

When searching for the set $\{J\}$ of cells overlapping with cell i, it is necessary to search bins b_l^r through b_r^r, plus bin 0. If a row is subdivided into very many bins, then the time to search a given bin is very short. However, the bucket will contain a large number of cells. On the other hand, if a row is subdivided into relatively few bins, the number of cells in each bin is relatively large, implying long search times since at least one large bin must be completely searched.

The goal is to find an expression yielding the optimal number of bins. It will be assumed that each row will have the same number of bins, that is, N_B bins per row. The length of each row r is defined as $l(r)$ and the sum of the lengths of the rows is represented by L. The determination of the optimal number of bins is based on the following: N_B is optimal when the average number of cells in each bin (c_b) is equal to the average number of cells in the bucket (c_0). On average, the number of cells in row r, expressed by $c_r(r)$, is proportional to the fractional length of row r relative to the total row length. That is,

$$c_r(r) = N_c \frac{l(r)}{L} \tag{3.9}$$

The model used to derive the optimal number of bins (N_B) assumes that the rows all have their left edge at the same x-coordinate. Further, the bins are uniformly distributed starting from the left edge of any row to the right edge of the longest row. Thus, the shorter rows have, possibly, one or more of their rightmost bins being empty. Hence, let $n_B(r)$ represent those number of bins actually used by a row r, implying that $N_B - n_B(r)$ bins for row r lie to the right of r and hence remain empty. Let x_R represents the rightmost edge among the right edges of all the rows. Then the number of bins in row r which are not empty is given by:

$$n_B(r) = N_B \frac{l(r)}{x_R} \tag{3.10}$$

Dividing the right-hand side of Eqn. 3.9 by the right-hand side of Eqn. 3.10 yields the average number of cells per actually used bin in row r:

$$c_b(r) = \frac{N_c}{N_B} \frac{x_R}{L} \tag{3.11}$$

The bins are implemented such that each bin boundary belongs to exactly one bin. Consider the case in which all the cells in a row are placed end-to-end starting from the left edge of the row, that is, it is assumed that there is no cell overlapping. In this case, exactly one cell touches each bin boundary. Then, the number of cells in the bucket for each row r is equal to the number of bin boundaries in each r. That is, if a cell intersects a bin boundary, one cell must go to the bucket, even if the edges from two different cells touch the same boundary. The latter is true since each of the two edges hashes to the same bin. Hence, a good approximation to the expected number of cells in the bucket is given by Eqn. 3.10. As stated above, the optimal number of bins is such that the expected number of cells per bucket equals the expected number of cells per bin. In other words, Eqn. 3.10 must equal Eqn. 3.11.

$$N_B^2 = N_c \frac{x_R^2}{L} \frac{1}{l(r)} \tag{3.12}$$

The square of the optimal number of bins is then obtained by averaging the right-hand side of Eqn. 3.12 over all of the rows.

$$N_B^2 = \frac{N_c \, x_R^2}{L \, N_r} \sum_{r=1}^{N_r} \frac{1}{l(r)} \tag{3.13}$$

This is the expression used by TimberWolfSC to determine the optimal number of bins per row.

Note that if all rows have the same length, that is, for all r:

$$l(r) = x_R = \frac{L}{N_r} \tag{3.14}$$

Then, by Eqn. 3.13, it follows that

$$N_B^2 = \frac{N_c \, x_R}{L} = \frac{N_c}{N_r} \tag{3.15}$$

Finally,

$$N_B = \sqrt{\frac{N_c}{N_r}} \tag{3.16}$$

From this equation, it is possible to determine the order of the number of bins. As mentioned previously, the order of the number of cells per row is $O\left(\sqrt{N_c}\right)$. It immediately follows that the order of the number of rows is given by:

$$O(N_r) = O\left(\frac{N_c}{\sqrt{N_c}}\right) = O\left(\sqrt{N_c}\right) \tag{3.17}$$

Hence,

$$O(N_B) = O(N_c^{0.25}) \tag{3.18}$$

The search time to determine the set of cells $\{J\}$ overlapping with a cell i is thus reduced by a factor of order one over N_B. Since the number of cells in a row grows according to:

$$c_r = O\left(\sqrt{N_c}\right) \tag{3.19}$$

the time complexity of the overlap search time is now

$$\frac{O\left(\sqrt{N_c}\right)}{O\left(N_c^{0.25}\right)} = O\left(N_c^{0.25}\right) \tag{3.20}$$

as a result of the bin storage technique. If the bin storage technique is not used, the time complexity of the average overlap search time would be given by $O\left(\sqrt{N_c}\right)$, where it is assumed that all of the cells in a row would have to be searched.

3.3.1.3 The third term in the cost function

The third term is the row length control penalty function, $C_3()$. Each row is assigned a penalty which is proportional to the absolute difference between the desired length of the row and the actual length of the row. The function $C_3()$ returns the following value:

$$C_3() = p_3 \sum_{r=1}^{N_r} \left| L_a(r) - L_d(r) \right| \tag{3.21}$$

where N_r is the total number of rows, and where $L_a(r)$ represents the actual (current) row length, that is, the sum of the widths of the cells currently placed in row r. Furthermore, $L_d(r)$ is the desired row length for row r. A parameter p_3 is used to normalize $C_3()$ for various grid sizes and expected values of row lengths. For the more than 20 circuits which were tested, each of which had a one micron (or less) grid size, $p_3 \approx 5$ was approximately the smallest value of p_3 which would yield uniform row lengths, without placing excessive emphasis on this term in the cost function. In the current implementation of TimberWolfSC, p_3 is an input parameter, with a default value of 5.

Upon the conclusion of the simulated annealing algorithm for stage 1, the necessary feed-through cells (if any) are inserted and all of the cells are compacted to the left. Since additional feed-through cells will usually be added to the rows, allowing some of the standard cells to occupy positions up to ten percent beyond the targeted length of their rows during stage 1 would be expected to yield realistic TEIC values. In this fashion, the standard cells have more row length in which to move around. This reduces the relative strength of the penalty function, and therefore produces better final values of the TEIC. The rows were assigned a length equal to 110 percent of the desired length of the longest row. The cells for a given row, therefore, had at least an additional 10 percent of length in which to move. However, the total cell length allowed in each row was still targeted for its corresponding desired length.

The penalty function $C_3()$ has a substantial impact on the performance of TimberWolfSC. The circuits used for the tests, courtesy of Hughes Aircraft, had approximately 800 cells. Table 3.1 presents the results for one of the circuits.

The average TEIL for a number of trials was compared with the method of Hughes Aircraft, which was largely a manual placement. Note that without $C_3()$, the average TEIL was reduced by 27 percent in comparison to Hughes. A fair amount of final row length modulation was noted for each of the 6 runs.

Penalty Function	No. of Trials	TEIL Reduction
None	6	27%
$C_3()$	6	33%

Table 3.1 Row length control penalty function - Example 1.

When $C_3()$ was added to the TimberWolfSC cost function, the results given in the last line of Table 3.1 were achieved. Note that this approach yielded an overall 6 percent *additional* reduction in the final TEIL. Furthermore, the control of the row lengths was excellent, a variation of only 2 or 3 percent, which is attributed to the wide variance in cell widths. Table 3.2 presents a similar comparison for another Hughes circuit.

Note that the additional average TEIL reduction was 5 percent over the absence of $C_3()$. The control of the final row lengths was again nearly precise. Thus, for both test cases, the addition of the penalty function of the form of Eqn. 3.21 not only yielded excellent control of the row lengths but also provided final TEIL values 5 to 6 percent lower than the case in which no such penalty function was used.

Penalty Function	No. of Trials	TEIL Reduction
None	4	24%
$C_3()$	2	29%

Table 3.2 Row length control penalty function - Example 2.

It would otherwise be expected that an additional penalty function would sacrifice some reduction in TEIL. However, it appears that without the row length control penalty function, the algorithm spends a significant portion of its time in the latter iterations (T values) trying to even out the rows, to reduce the overlap penalty, and less time toward minimizing the TEIL. The addition of $C_3()$ appears to keep the value of the overlap penalty somewhat smaller than it would otherwise be, therefore permitting more emphasis on TEIL reduction in the latter iterations. Without the use of $C_3()$, the number of attempted new configurations per value of T must be made very large in order for the final row lengths to be uniform. On the other hand, the use of the penalty function results in uniform row lengths with substantially fewer attempted new configurations per cell per value of T.

Tests were also conducted on a second formulation for the function $C_3()$. This expression is shown below:

$$C_3() = p_3 \sum_{r=1}^{N_r} | L_a(r) - L_d(r) |^2 \tag{3.22}$$

In this case the penalty is proportional to the square of the absolute difference between the actual and desired row lengths. For the same Hughes Aircraft circuit which was used to obtain the results in Table 3.1, the average TEIL was reduced by 24 percent in comparison to Hughes. For this penalty function, the row lengths were *perfectly* uniform. While this result may be more desirable (because of the perfect row length control) than the no penalty function result, despite the sacrifice of three percent in TEIL, it does not compare well with the penalty function of Eqn. 3.21 which yielded a 33 percent reduction.

3.3.2 An alternative objective function

Tests were conducted on a second form for the primary objective function, designated as $C_1'()$. Although there is a distinct correlation between the final chip area and the final TEIL, the relationship on a per trial basis is characterized by some variance. Since chip area is related closely to the congestion in the channels, a new objective function was pursued with the goal of measuring more closely the local channel congestion. The best possible objective would be to complete the global routing after the generation of each new configuration. However, the computation time required for a 5000-cell circuit in which 10^8 new configurations require evaluation would be prohibitive. A new objective function was designed in which the density of each channel was *approximated*. This objective function was implemented as follows: Vertical *cut lines* were established in each horizontal channel as shown in Figure 3.5. The spacing of the cuts was approximately one-fourth of an average cell width. The number of cut lines in the horizontal direction is represented by N_{ct}. For each channel *ch*, a histogram *h[ch]* is maintained, in which the histogram value *h[ch][c]* indicates the number of nets crossing cut *c* in that channel.

For each channel *ch*, the cut line within *ch* having the greatest number of nets crossing it determines the anticipated density of *ch*. The sum of the anticipated densities is then a measure of the variable portion of the chip height. This *estimated* sum is designated D_e.

$$D_e = \sum_{ch=1}^{N_{ch}} \max_{1 \le c \le N_{ct}} h[ch][c] \tag{3.23}$$

The span of a net n in a channel ch bordered from below by row r_{i-1} and bordered from above by row r_i is determined as follows: (1) Starting from row r_i and counting upward, that is, $r_i, r_{i+1}, r_{i+2}, \cdots$, find the nearest row having pins belonging to net n. Designate this row as r_T. (2) Next, starting from row r_{i-1} and counting downward, that is, r_{i-1}, $r_{i-2}, r_{i-3}, \cdots$, find the nearest row having pins belonging to net n. Designate this row as r_B. (3) Let L_r^n be the location of the leftmost pin of net n which is on either row r_T or r_B. (4) Let R_r^n be the location of the rightmost pin of net n which is on either row r_T or r_B. The values of L_r^n and R_r^n constitute the span of net n in row r. If net n has pins only on row r_B, then L_r^n and R_r^n constitute the span of net n in channel *ch*. That is, it is arbitrarily assumed that n will be routed in the channel above row r_B. For all other cases, if either r_T or r_B is null, then net n does not have a span in *ch*. This is indicated by setting L_r^n and R_r^n to null.

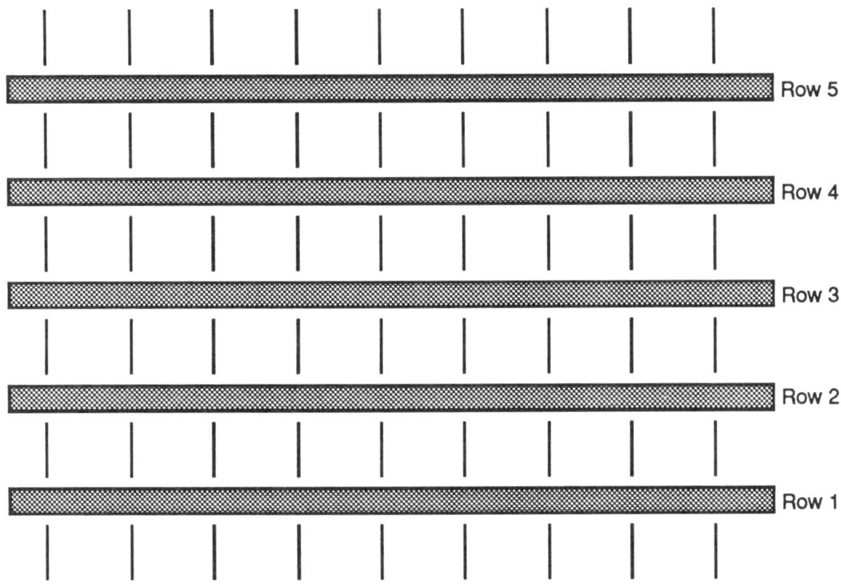

Figure 3.5 Vertical cut lines for the new objective function.

The histogram values for the cut lines are determined as follows: (1) For each net n, determine the list of channels for which L_r^n and R_r^n are not null. (2) Let $h_x[ch][c]$ represent the x-coordinate of the cut line c of channel ch. Then add one to each histogram such that:

$$L_r^n \leq h_x[ch][c] \leq R_r^n \tag{3.24}$$

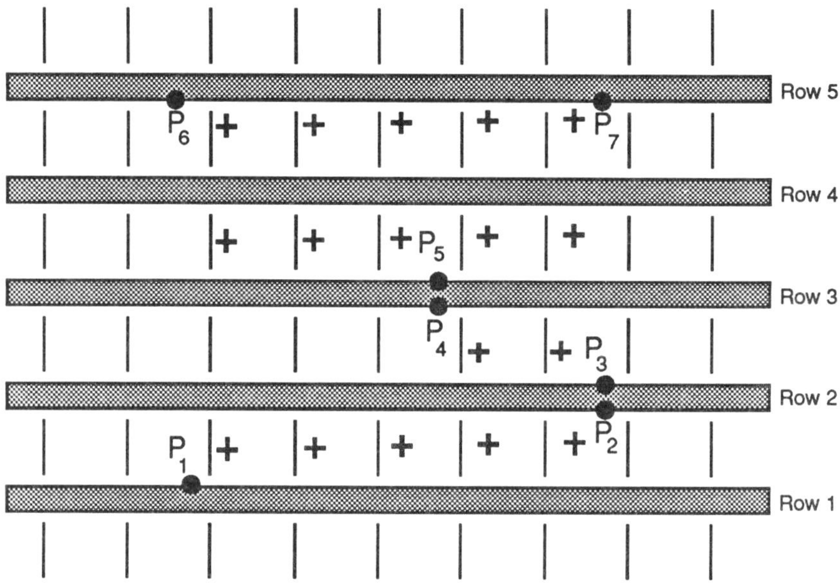

Figure 3.6 Those cut lines marked with a "+" are incremented by one as a result of this seven-pin net.

An example of a seven-pin net is shown in Figure 3.6. Those cut lines marked with a + are incremented by one as a result of this net. Note that row 4 in this example does not have a pin belonging to the net. Nonetheless, the formulation of the above algorithm attributes net crossings to both the cut lines in the channel above and the channel below row 4. This is effectively a *penalty*, which is assigned because a feed-through cell instance would have to be inserted into row 4 in order to complete the detailed routing. This penalty discourages the use of feed-through cells which contribute to row, and hence chip, width.

The actual objective function $C_1'()$ which was tested is shown below:

$$C_1'() = \sum_{ch=1}^{N_{ch}} \sum_{c=1}^{N_{ct}} h[ch][c]^2 \qquad (3.25)$$

The performance of $C_1'()$ was compared to that of the function $C_1()$. The results shown in Table 3.3 are for a double-metal-technology circuit from Intel. Note that the four trials with the objective function $C_1()$ had an average of $D_e = 223$. Meanwhile, the new objective function $C_1'()$ yielded an average of $D_e = 166$. The 26 percent reduction in the estimated total channel density represents a sizable improvement.

Objective Function	D_e
$C_1()$	226
$C_1()$	214
$C_1()$	220
$C_1()$	235
$C_1'()$	166

Table 3.3 New objective function – Intel BLK12 example.

The evaluation of $C_1'()$ required substantially more CPU time than $C_1()$. In fact, the difference was slightly more than one order of magnitude. This was partially due to a somewhat inefficient implementation of the evaluation of $C_1'()$. Nonetheless, there is much more analysis to be done with respect to the new objective function. For each new configuration, the spans of the affected nets must be broken down channel by channel and all of the relevant cut lines must be updated. Furthermore, during the first half of stage 1, when many of the new configurations are generated by large moves, it often arises that a substantial fraction of the cut lines must be updated.

The example shown in Table 3.4 is also from Intel. Note that the six trials with $C_1()$ yielded an average of $D_e = 204$. Meanwhile, the new $C_1'()$ yielded $D_e = 136$. The 33 percent reduction in the worst-case track count again represents a sizable improvement. It is clear from Tables 3.3 and 3.4 that $C_1'()$ produces a value of D_e which is on the order of 30 percent smaller than that produced by $C_1()$. Additional measurements were then taken to determine the relationship between the minimization of D_e and the minimization of chip area.

Objective Function	D_e
$C_1()$	196
$C_1()$	190
$C_1()$	213
$C_1()$	217
$C_1()$	216
$C_1()$	194
$C_1'()$	136

Table 3.4 New objective function – Intel BLK1 example.

Table 3.5 shows the results for a 100-cell Intel example. The first row of the table shows the data for their manual placement and automatic routing. The second row in the table shows the data produced by the objective function $C_1()$. Note that D_e was reduced by 27 percent in comparison to the manual placement. The third row shows the data for the new objective function. Note that D_e was reduced now by 42 percent. However, the final chip area (after detailed routing) was exactly the same as that in which $C_1()$ was used as the objective function. Since the routing was performed by Intel, additional data was not available. One can surmise that the unexpected results are due to: (1) The estimation scheme fails to account for electrically equivalent pins, which has a significant effect on the total channel density as presented later. (2) The *penalty* for a required feed-through path, for example the fourth row in Figure 3.6, applied by charging the cut lines above and below the rows may be too strong.

Objective Function (or Manual)	D_e	Chip Area
Manual	105	90×81
$C_1()$	77	88×77
$C_1'()$	61	88×77

Table 3.5 New objective function – 100-cell Intel example.

The results of another test, a 180-cell Gould-AMI example, are shown in Table 3.6. The first row of the table shows the data for $C_1()$, where D_e^0 represents the estimated density as a result of the Gould-AMI placement. Note that D_e was reduced by 43 percent in comparison to the automatic placement of Gould-AMI. The actual number of detailed routing tracks required was 78. The second row of Table 3.6 shows the data for $C_1'()$. Note that D_e was reduced by 12 percent less than for the case in which $C_1()$ was used. The new objective function resulted in a 53 percent reduction in D_e in comparison to the Gould-AMI placement, which represents an additional reduction of 10 percent over the original objective function. Despite having a D_e value which was smaller by 19, the actual number of tracks was actually six more for the new objective function.

Objective Function	TEIL Reduction	$\frac{D_e}{D_e^0}$	Tracks
$C_1()$	30%	108 / 188	78
$C_1'()$	18%	89 / 188	84

Table 3.6 New objective function – 180-cell Gould-AMI example.

The data for a third example is shown in Table 3.7, in which $C_1'()$ actually resulted in a substantial chip area reduction. Note that the three trials with the $C_1()$ objective function yielded an average of $D_e = 181$. Meanwhile, the new objective function yielded $D_e = 108$, which represents a 41 percent reduction. In fact, after detailed routing, the height of the placement yielded by the new objective function was 23 percent smaller. Since the final chip widths were about the same for both placements, the new objective function yielded a 23 percent area reduction for this example.

The conclusions that were reached following the comparisons of the two objective functions were: (1) The CPU time required to evaluate $C_1'()$ was slightly over an order of magnitude longer than for the evaluation of $C_1()$. (2) The final-chip-area comparisons yielded mixed results. (3) The capability of weighting individual nets is largely lost with the implementation of the $C_1'()$ objective function.

Objective Function	D_e	Chip Height
$C_1()$	177	Not Recorded
$C_1()$	181	1219
$C_1()$	184	Not Recorded
$C_1'()$	108	949

Table 3.7 New objective function – Intel P example.

The loss mentioned in (3) above is rather crucial. The ability to optimize performance as well as to minimize area is the true objective. With respect to point (1) above, a better implementation could probably cut the CPU time penalty down to a factor of three or so. Also, a parallel algorithm could be designed which would dramatically cut down on the evaluation time. For example, for each net the cut lines in each channel could be updated simultaneously.

The reason for the mixed final chip area results is probably due to over-estimation of the channel densities. Table 3.8 presents comparisons for ten TimberWolfSC placements in which the estimated total channel density D_e was compared with the actual total channel density D, obtained with the algorithms presented in this chapter.

From the table, note that D_e is on average about 33 percent larger than D. Certainly one could try to reduce the penalty for a required feed-through path and thereby attempt to approximate better the true density. However, this would increase the CPU time even more, and possibly by a significant amount. Furthermore, since the circuits in Table 3.8 required relatively few feed throughs, it is my opinion that the fact that the implemented scheme fails to account for electrically equivalent pins is the major reason for the discrepancy shown in Table 3.8. Later on, the plausibility of this argument will be brought out when the TimberWolfSC global router results are presented.

Circuit	D	D_e	Percent Deviation
EX1	651	781	20
EX1	417	562	34
EX2	109	150	37
EX2	121	169	39
EX3	102	130	27
EX3	96	135	29
EX3	90	131	40
EX4	71	85	19
EX4	51	69	35
EX5	289	434	50
Avg.			33

Table 3.8 Comparison of D and D_e for 10 industrial circuits.

3.3.3 The generation of new states function

Details of the function responsible for the generation of new configurations for stage 1 are now presented. Tests have corroborated the theoretical results which have shown that the initial configuration has no influence on the final value of the TEIC.[1] Hence it is common to begin with a random initial placement of the standard cells. The function generate() produces the next new configuration by making a weighted random selection from one of the following: (1) the displacement of a randomly chosen single cell to a randomly chosen new location, on a randomly chosen row, (2) the interchange of two randomly selected cells, or (3) an orientation change for a cell. The selection of alternative 3 occurs only if alternative 1 resulted in a new configuration being rejected.

In the simulated annealing algorithm, each pass through the body of the *inner loop* begins with the selection of either alternative 1 or 2 above. The ratio r of single cell displacements to pairwise interchanges can have an important effect on the final TEIL. The normalized final TEIL for a variety of circuits (courtesy of Gould-AMI Semiconductors) versus the ratio r of single-cell displacements to pairwise interchanges is shown in Figure 3.7. For all of the circuits, the data was taken for the following values of r: 0.5, 1, 2, 4, 8, and 16. The fitted curve in Figure 3.7 passes through the measured (normalized) TEIL values. The *inner loop* consisted of 200 attempted new configurations per cell (or a total of 300,000 new configurations). It is apparent from this figure that values of r in the range from 3 through 8 yield TEIL values which are within one percent of the minimum. This wide range gives robustness to the algorithm.

[1] D. Mitra, F. Romeo, and A. Sangiovanni-Vincentelli, "Convergence and Finite-Time Behavior of Simulated Annealing," (1985).

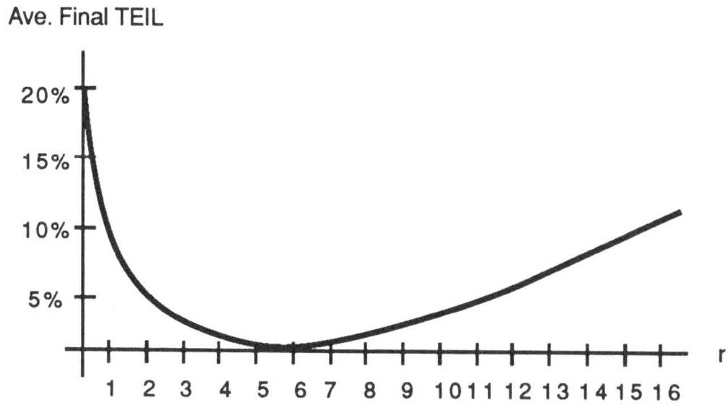

Figure 3.7 The normalized final TEIL for a variety of circuits versus the ratio *r*.

If a single cell displacement attempt failed to gain acceptance, the next new configuration generated is an orientation change (a mirror of the x coordinates of the cell). The number of times such new configurations are proposed was arbitrarily limited to 10 percent of the number of times the opportunity arises for an orientation change of the cell. This was done to limit, and in fact, to make negligible the increase in CPU time due to permitting cell orientation changes. Tests on more than 10 circuits failed to show any effect on the quality of the final TEIL. Since it is usually the case that 100 to 1000 new configurations per cell are attempted per value of T at the low values of T, 10 to 100 orientation changes per cell are usually attempted. It is therefore not surprising that this number of orientation changes per cell per T appears to be adequate to explore this aspect of the configuration space.

Note that orientation changes are not attempted following an acceptance of a single cell displacement. If the current placement of a cell was such that a comparatively large move (in terms of the potential change in cost) was needed, then it was assumed that a relatively small refinement is unnecessary at this time. Table 3.9 indicates the final total estimated interconnect costs for eight trials in which orientation changes (mirroring of the x coordinates of the cells) were allowed versus eight trials when such orientation changes were not allowed. Note that permitting the cell orientation changes resulted in an average TEIL which was three percent smaller.

No. of Trials	Average with Orientation Changes	Average without Orientation Changes
8	113,900	117,900

Table 3.9 Average final cost with and without orientation changes.

3.3.4 The inner loop criterion

The inner loop criterion in TimberWolfSC, that is, the number of iterations of the inner loop, is specified in terms of the number of attempted new configurations per cell per value of T (designated as A_c). The number of iterations of the inner loop for a given value of T is then given by:

$$A = A_c \cdot N_c \tag{3.26}$$

where N_c is the number of standard cells.

The effect of the inner loop criterion was noted for several 1500-cell circuits, each of which was also placed by an industrial placement program (Gould-AMI's CIPAR or by Hughes Aircraft). In Figure 3.8, a plot of the average TEIL reduction (in comparison to the industrial placements) versus A_c for these circuits is shown. Note that for circuits of this size, $A_c = 400$ is sufficient to yield the best results. Also note that reasonably good results can be obtained for substantially smaller values of A_c.

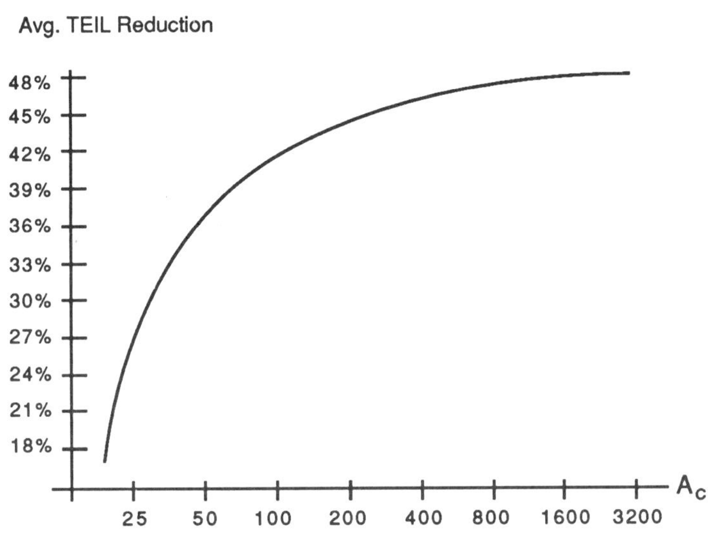

Figure 3.8 The average TEIL reduction, in comparison to the industrial placements, versus A_c. Twenty trials were performed for A_c = 12, 25, 50, and 100. Twelve trials were performed for A_c = 200, four trials for 400, two trials for 800, and one trial for 1600 and 3200.

The execution time for stage 1 of TimberWolfSC is directly proportional to A_c. The CPU time for a typical $A_c = 400$ run was about 4 hours on an IBM 3081K. Note that for $A_c = 25$, the reduction in TEIL was 15 percent less than the best results. However, the CPU time in this latter case was 16 times less than that for the $A_c = 400$ tests. While the $A_c = 400$ case represents a very substantial amount of CPU time, it has been observed that the additional 15 percent reduction in TEIL often yields an additional final chip area reduction of as much as 10 to 15 percent. From this standpoint, in the latter stages of the circuit design process, the additional CPU time is usually justified. On the other hand, in the early stages of the design, it is best to choose a smaller value of A_c.

3.3.5 The range limiter

The generate() function in TimberWolfSC is controlled by a *range limiter.* Large-distance moves usually imply large values of ΔC. At low temperatures, only moves which approximately satisfy $\Delta C \leq 0$ have a reasonable chance of being accepted. Hence, at low temperatures, the large-distance moves are almost invariably rejected. In order to generate moves which have a reasonable probability of acceptance, these large-distance moves are therefore prohibited by the use of a range-limiter window. When a cell is selected for displacement, such as the cell in the center of the third row in Figure 3.9, the randomly selected new location for the cell: (1) must be in a row whose center in the y-direction lies within the range-limiter window centered on the current position for the cell, and (2) must have an x-coordinate which lies within the window. When a cell i is selected to participate in a pairwise interchange of cells, a second cell j (to be interchanged with cell i) is selected such that: (1) the vertical distance between the centers of the cells is less than or equal to the vertical span of the range-limiter window and (2) the horizontal distance between the centers of the cells is less than or equal to the horizontal span of the window.

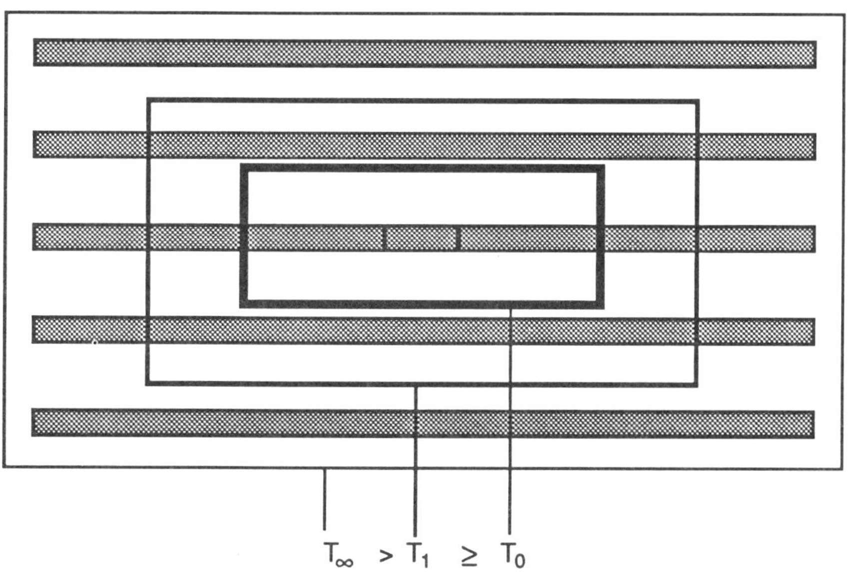

Figure 3.9 The span of the range-limiter window is proportional to the log(T). At $T = T_\infty$, the window has maximum span, and at $T = T_0$ interrow cell moves are prohibited.

At the beginning of stage 1, when T is at its maximum value (indicated as T_∞ in Figure 3.9), the window extends beyond the entire chip, thereby allowing moves of maximum distance. Since six or seven orders of magnitude of T are used, the range limiter span is reduced proportional to the logarithm of T. In Figure 3.9, the span of the window is indicated for an intermediate value of $T = T_1$. As T approaches the value $T = T_0$, the vertical span of the window is such that interrow cell moves are no longer permitted. This condition marks the end of stage 1 of TimberWolfSC.

The actual formulas controlling the respective window dimensions are shown below:

$$W_x(T) = W_x^{\infty} \left\{ \varepsilon \frac{\log(T)}{\log(T_{\infty})} \right\} \tag{3.27}$$

$$W_y(T) = W_y^{\infty} \left\{ \varepsilon \frac{\log(T)}{\log(T_{\infty})} \right\} \tag{3.28}$$

where W_x^{∞} represents the horizontal span of the chip and W_y^{∞} represents the vertical span of the chip.

Several experiments were performed to determine the most appropriate value for ε. The specific results of one of these experiments are shown in Table 3.10, where the data was for a 1500-cell circuit from Gould-AMI. The net list data was specified with a one micron grid size. The value of A_c was set to 200. Since there was little difference in the final values of the TEIL, $\varepsilon = 0.2$ was chosen due to the fact that the number of iterations to satisfy the *stopping criterion* was minimized.

Denote d_s as the row separation distance. Then, in accordance with Figure 3.9, $T = T_0$ is such that:

$$W_y(T_0) = d_s = W_y^{\infty} \left\{ \varepsilon \frac{\log(T_0)}{\log(T_{\infty})} \right\} \tag{3.29}$$

ε	Final TEIL	Iterations
200	674,524	157
200	715,421	158
20	677,881	149
20	705,793	150
2	674,476	141
2	694,969	141
0.2	683,851	140
0.2	680,318	140

Table 3.10 The effect of ε on the range limiter.

Note that if the row separation distance is very large, then the range limiter will prevent cells from moving from one row to another at a relatively high value of T. This is not desirable since many of the moves being accepted at a relatively high T actually increase the objective function. Based on experimentation with 1500-cell circuits with a grid size of one micron, the best performance was obtained with $\varepsilon = 0.2$ where d_s equaled the height of the rows. For these circuits, the onset of the prevention of interrow exchanges occurred for $T = T_0 \approx 5.0$. In order to obtain the best performance from stage 1 of TimberWolfSC, it was necessary to normalize ε, for a circuit of any number of cells, any grid size, and for any user-specified row separation (d_s).

In order to normalize ε as indicated above, it is first necessary to normalize the value of T_∞ with respect to grid size of the input data. This is apparent from Eqns. 3.23, 3.24 and 3.25, where functions $C_1()$ and $C_3()$ return values which are proportional to grid size, and where function $C_2()$ returns a value which is quadratic with grid size. However, it is also necessary to take into account the total number of cells N_c in the circuit, since the total value of the objective function is strongly affected by N_c. A good method for normalizing T_∞ is to sample the cost values at essentially $T = \infty$ and to take the average of all of the positive ΔC values ($\overline{\Delta C}$). Then, select

$$T_\infty = \kappa \overline{\Delta C}$$

where κ is a small integer.

On the other hand, a simple method which has worked well in tests on more than 30 industrial circuits is to account for both grid size and N_c with the following expression:

$$W_c = \sum_{c=1}^{N_c} w_c(c) \tag{3.30}$$

where $w_c(c)$ returns the width of cell c. Defining W_c^∞ as W_c for a typical 2500-cell circuit having a grid size of one micron, the normalization of T_∞ proceeds as below:

$$T_\infty = \frac{W_c}{W_c^*} T_\infty^* \tag{3.31}$$

The value of T_∞^* for a typical circuit of 2500 cells and one micron grid size yielding the best performance was on the order of 10^6.

The value of T indicating the onset of the prevention of interrow cell interchanges (and the end of stage 1), that is, T_0, also requires normalization. Using the same procedure leading to the development of Eqn. 3.31, with $T_0^* = 5.0$, it follows that:

$$T_0 = \frac{W_c}{W_c^*} T_0^* \tag{3.32}$$

With the normalization of T_∞ and T_0, completed, it is then possible to normalize ε. Solving Eqn. 3.28 for ε:

$$\varepsilon = \frac{d_s}{W_y^\infty} \frac{\log(T_\infty)}{\log(T_0)} \tag{3.33}$$

where T_∞ is given by Eqn. 3.31 and where T_0 is given by Eqn. 3.32. Equation 3.33 specifies the normalized value of ε for any circuit whose total cell length is reflected in the computation of the values of T_∞ and T_0, and whose row separation distance is given by d_s.

3.3.6 The control of T

The function update(T) is expressed by

$$T_n = T \cdot \alpha(T), \quad 0 < \alpha(T) < 1 \tag{3.34}$$

where T_n is the new value of the temperature derived from the current value T. The cooling schedule used in TimberWolfSC was determined experimentally, with the following observations: (1) Three to five iterations are performed in which virtually every new configuration was accepted and where T is reduced quite rapidly from iteration to iteration. (2) After having left the high T regime, the reduction in T proceeded in such a manner that ΔC is approximately the same from iteration to iteration. (3) When T is reduced below 1.5, T is reduced very rapidly so as to firmly converge the cost function.

For a typical circuit of 2500 cells, specified with a grid size of one micron, Table 3.11 contains the data for α as a function of T for which the best results were obtained with TimberWolfSC. The entries in the first column indicate the smallest value of T for which α returns the corresponding entry in the second column. That is, for $T \geq 40{,}000$, α is set to 0.80, for $20{,}000 \leq T < 40{,}000$, α is set to 0.84, and so on.

For $T \geq$	α
40,000	0.80
20,000	0.84
10,000	0.88
5,000	0.91
200	0.94
100	0.90
50	0.85
5	0.80
1.5	0.70
0	0.10

Table 3.11 The TimberWolfSC cooling schedule, α vs. T.

It was arbitrarily determined that approximately 120 temperature values were to be considered in a typical execution of the simulated annealing algorithm. Based on extensive experimentation for circuits of this size and gridsize, the range of T from 200 to 5000 influenced most strongly the performance of TimberWolfSC. Consequently, as many temperature points as possible were placed in this range while arbitrarily electing not to reduce T more rapidly than by a factor of 0.80, until the very end. The performance of TimberWolfSC did not appear to be very sensitive to the temperature profile outside the crucial range of 200 to 5000.

The cooling schedule defined by Table 3.11 also requires normalization for circuits of various numbers of cells and for circuits of various specified grid sizes. The normalization factor used to derive Eqn. 3.31 is also used to normalize any circuit to the α vs. T data of Table 3.11. That is, instead of using $\alpha(T)$, a value $\alpha(T')$ is used where

$$T' = \frac{W_c}{W_c^*} T \tag{3.35}$$

3.3.7 The effects of net weighting

This subsection presents the effect of net-weighting on the results of stage 1 of TimberWolfSC. Table 3.12 has data for an 800-cell Hughes Aircraft circuit in which the initial TEIL was 1,996,252 as a result of their manual placement. The value of A_c was set to 400. The third column in the table represents the total number of feed-through cells which had to be added in order to complete the detailed routing. The fourth column represents the horizontal, or x-directed, component of the TEIL (H-TEIL). The row separation distance was the same for the three trials shown in the table.

H Weight	V Weight	No. Feeds	H-TEIL
1.0	1.0	365	677,341
1.0	0.7	446	628,061
1.0	1.5	341	726,852

Table 3.12 Effects of net weighting.

Note that if the vertical weighting factor is reduced relative to the horizontal weighting factor (row 2 in Table 3.12) then the H-TEIL is reduced as expected. If a greater percentage of the TEIL is in the vertical direction, as it is for this case, then it would be expected that the number of feed-through cells required would increase. In fact, the second row of Table 3.12 confirms this. Similarly, if the vertical weighting factor is increased relative to the horizontal weighting factor (row 3 in Table 3.12) then the H-TEIL is increased as expected. If a smaller percentage of the TEIL is in the vertical direction, as it is for this case, then it would be expected that the number of feed-through cells required would decrease. This again is confirmed by the third row of Table 3.12.

Hence, there is an inverse relationship between the number of feed-through cells required and the H-TEIL. This can be viewed as a trade off of horizontal chip width (the effect of the feed throughs) versus vertical chip height (the effect of additional wiring in the horizontal channels). As an example, for double-metal technology, most of the required feed-through paths can be accomplished by either the use of the built-in feeds of the standard cells, or by simply routing over the rows. In this case it would be wise to trade off H-TEIL for the additional (effective) feed-through path requirements. That is, the latter will not increase the chip width for double metal technology and the former will reduce chip height.

Table 3.13 has data for a 2700-cell, double-metal Gould-AMI circuit in which the original chip area was 691 × 691 mils^2 as a result of their automatic placement and routing. The row separation distance equaled the height of the rows.

H Weight	V Weight	Final Area (mils^2)	Notes
1.0	0.1	500 × 500	$A_c = 600$
1.0	0.4	420 × 420	$A_c = 600$
1.0	0.3	400 × 400	$A_c = 600$
1.0	0.3	379 × 379	$A_c = 800$

Table 3.13 Effects of net weighting on a double metal circuit.

The first row of Table 3.13 represents a TimberWolfSC placement in which the vertical net weighting factor was 0.1 times the horizontal factor. The chip area reduction was 48 percent. However, tens of thousands of feed-through paths were required, and not nearly that many built-in feed throughs were available. Consequently, on the order of 10 thousand feed-through cells had to be added, which added a considerable amount to the final chip width. Also, when a feed-through path was required in a row after most of the built-in feed throughs were exhausted for that row, the search for the remaining paths resulted in very long horizontal wire runs.

The vertical net weighting factor was increased to 0.4 of the horizontal for the next trial as shown in the second row. This relative weighting scheme resulted in a final chip area which was a 63 percent reduction in comparison to the Gould-AMI placement. For the third trial, the vertical net weighting factor was reduced to 0.3. Note that the final chip area reduction was also more impressive, now 66 percent.

The final run was repeated with the vertical net weighting factor of 0.3, however, A_c was increased to 800. This resulted in the best placement obtained for this circuit. The final chip area reduction was 70 percent over the AMI placement and the number of actual feed-through cells which had to be inserted decreased from 13,789 to 1925.

3.4 The Algorithms for Stage 2 of TimberWolfSC

This section describes the algorithms for stage 2 of TimberWolfSC, which includes the global router. When T is reduced to the point such that $T = T_0$, the vertical span of the range-limiter window, expressed by $W_y(T_0)$ has been reduced to an amount less than the center to center spacing between the rows. Hence, at the point in which $T = T_0$, interrow cell moves are no longer permitted. At this point, stage 1 ends and TimberWolfSC begins a transition to stage 2.

Since the cells are constrained to their current rows, the feed-through path requirements are satisfied before continuing on to stage 2. This strategy permits the adjustment of feed-through positions during the last two stages of placement optimization. Depending on the technology, the feed-through-path requirements are satisfied in one or more of the following three ways: (1) Feed-through-cell instances are inserted into the rows, for example, into the second row shown on the left side of Figure 3.10. (2) Built-into-the-cell feed-through paths are utilized to effectively pass *through* a row, for example, through the cell in the second row of the center portion of Figure 3.10. (3) Indiscriminate routing over the rows, for example, over the second row as shown on the right side of Figure 3.10. In the current implementation of the third method, the permitted cross-over points on each cell must be specified.

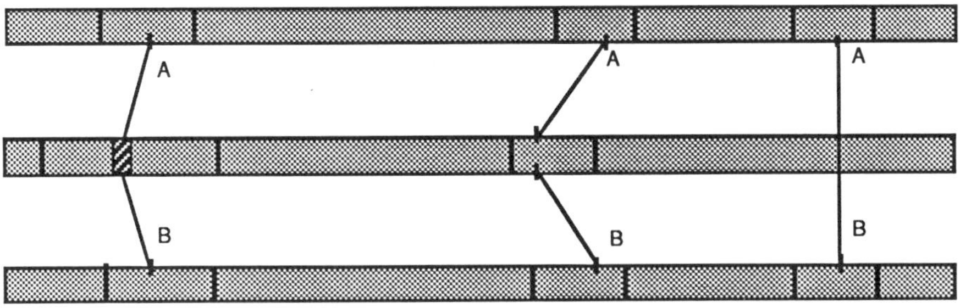

Figure 3.10 Feed-through paths can be implemented by (1) inserting feed-through cell instances, as in the second row on the left, (2) using built-in-to-the cell feeds, as in the middle, and (3) by routing over the row, as on the right.

The transition to stage 2 has another facet. As $T \rightarrow T_0$, the penalty function $C_2()$ (expressed by Eqn. 3.8) has eliminated most instances of cell overlapping. However, it is often the case that there are a few instances of cells overlapping by a small amount at the end of stage 1. Prior to beginning stage 2, any residual instances of overlapping cells are removed. The cells for each row are sorted according to the location of the x-coordinate of their centers. The cells for each row are then *replaced* side-by-side starting from the left edge of the respective row. Therefore, the cells are tightly packed starting from the left edge of each row, with no overlapping and no empty space between the cells. Measurements have indicated that the TEIL changes by less than one percent after this procedure.

There is a reason for permitting residual cell overlapping at the end of stage 1. If the relative *strength* of the penalty function $C_2()$ is large compared to the TEIL term $C_1()$, that is, if the normalization parameter p_2 of Eqn. 3.8 is large in a relative sense, the simulated annealing algorithm pays relatively more attention toward reducing the value of the penalty function and relatively less attention toward minimizing the TEIL. On the other hand, if p_2 is made smaller, the simulated annealing algorithm pays relatively more attention toward minimizing the TEIL, and relatively less attention toward reducing the value of the penalty function. The best results have been obtained when p_2 is made as small as possible (permitting as much residual overlap as possible) without experiencing a noticeable (for example, more than one percent) change in the TEIL after the *infeasibilities* (overlaps) have been removed by the procedure outlined above. That is, the *maximum* possible attention is paid to minimizing the term of choice, namely $C_1()$, in the overall objective function.

Stage 2 continues the simulated annealing placement optimization based on TEIC from $T = T_0$ down to $T \rightarrow 0$. Furthermore, the second stage comes to a close with the execution of the global router. The remainder of this section is devoted to the implementation of the stage 2 simulated annealing functions and to the global router algorithm.

3.4.1 Implementation of the stage 2 simulated annealing functions

The standard cells are not permitted to change rows during stage 2. Hence, the third term in the objective function, the row length control penalty function, $C_3()$, is not necessary. Effectively, in Eqn. 3.21, $p_3 = 0$ is maintained during stage 2. Furthermore, as stated above, there are no instances of cell overlapping upon entering the simulated annealing algorithm of stage 2. As will be seen, the new configurations are generated in such a manner that an instance of cell overlapping is never created. Therefore there is also no need for the overlap penalty function $C_2()$. Effectively then, in Eqn. 3.8, $p_2 = 0$ is maintained during stage 2. The cost function for stage 2 simulated annealing thus consists of only the term $C_1()$ (the TEIC) expressed by Eqn. 3.7.

Since the horizontal span of the range-limiter window was also made very small as $T \rightarrow T_0$, the generation of new configurations is very local in stage 2. In fact, the function generate() takes on a substantially different form because interrow cell moves are not permitted for $T < T_0$. For every iteration of the inner loop, five steps are completed. (1) Randomly select a cell i from the set of all standard cells and feed-through cells. (2) Find the left neighbor, L, of selected cell i, if such a neighbor exists. (3) Find the right neighbor, R, of cell i, if such a neighbor exists. (4) If L is null, then the next new configuration attempted is an interchange of cells i and R. On the other hand, if R is null, then the next new configuration attempted is an interchange of cells i and L. If neither L nor R is null, one of them is randomly selected to be interchanged with i. If this new configuration is not accepted, then the next new configuration is generated by interchanging i with the other neighbor. (5) The next new configuration is generated by attempting an orientation change for cell i.

The neighboring pairs of cells are interchanged in such a manner as to not create any overlap, nor to create any empty space between cells. This is in contrast to the stage 1 generate() function which interchanges the locations of the geometric centers of the pair of cells. The *stopping criterion* is satisfied when three successive inner-loop iterations end with the same value of the TEIC.

3.4.2 The first phase of the global router

The minimization of the TEIC (or the TEIL) does not necessarily imply minimum chip area. However, it is very important to minimize the TEIL for circuit performance reasons. The strategy used in TimberWolfSC is to first minimize the TEIL, and *then* to minimize the chip area. Because interrow cell exchanges are no longer attempted, the lengths of the respective rows are fixed. Since the width of the chip (or at least its core portion) is determined by the longest row, minimizing the chip area implies minimizing the height of the chip. The variable portion of the chip height is the number of wiring tracks required to complete the routing of the chip. A good approximation to the number of wiring tracks required (as will be shown in the results section) is the total channel density, or D, expressed as follows:

$$D = \sum_{ch=1}^{N_{ch}} d(ch) \tag{3.36}$$

where the function *d()* returns the density of channel *ch* and where N_{ch} is the total number of channels. The objective, therefore, is to minimize the sum of the densities of all of the horizontal channels. In order to compute D, it is necessary to perform a global routing step.

The global router has two phases. The first phase is executed on each net individually. The purpose of this phase is to generate many possible routes for each net. The second phase minimizes the total channel density by simultaneously considering the routes of all nets, hence eliminating the classical routing-order-dependence problem. This latter phase examines the alternative routes for each net, as generated in the first phase. The remainder of this subsection presents additional algorithmic details on the first phase. The next subsection will examine the second phase in greater detail.

Standard cells are usually designed with electrically equivalent (internally connected) groups, or *clusters*, of pins. The pin clusters usually have pins on both the top and bottom sides of the cell. The location of a pin cluster in the x direction is taken to be the average of the x locations of the constituent pins. A portion of a net which must connect two pin clusters is referred to as a *net segment*. It often arises that a pin cluster from one cell must be connected to a pin cluster from another cell on the same row. If each cluster has a pin on the top of the cell as well as a pin on the bottom of the cell, then the net segment connecting the two clusters is referred to as a *switchable net segment*. A switchable net segment can be implemented by either a net segment in the channel below the row or a net segment in the channel above the row. The channel assignments for the switchable net segments are made based on the minimization of Eqn. 3.36.

The single net illustrated in Figure 3.11 requires the interconnection of six pin clusters on six different cells. In Figure 3.11, the second row from the bottom contains three pin clusters for this net. Note that the two leftmost clusters of these three can be interconnected by a net segment in either the channel below the row (solid line), or in the channel above the row (dotted line).

The graph of possible interconnections can, in general, have cycles. For example, consider the following four pin clusters in Figure 3.11: the cluster in the first row, the third row and the two rightmost clusters in the second row. One of the purposes of phase 1 is to eliminate redundant edges in a cycle, for example, the dotted edge in the aforementioned cycle in Figure 3.11. Note that the global router only routes interconnections around the outside of the rows if feed-through insertion was not permitted.

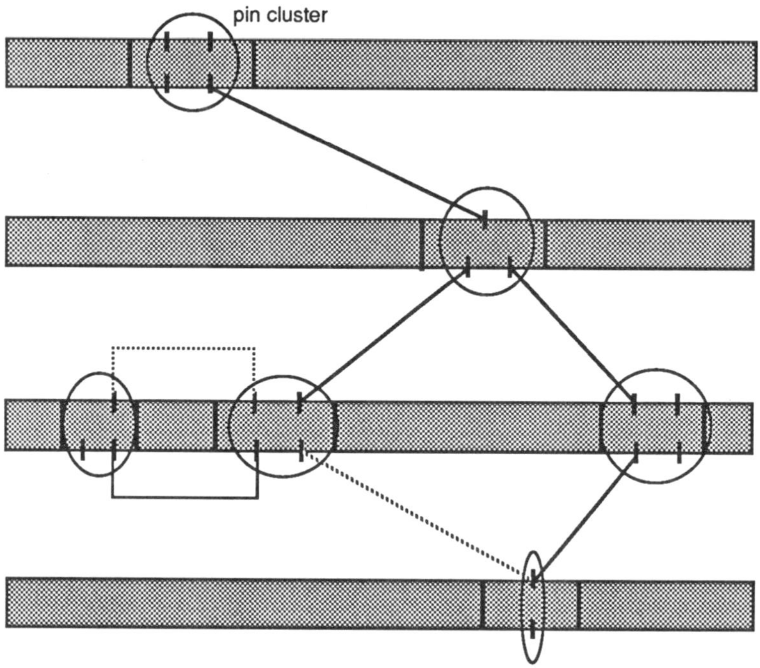

Figure 3.11 The graph of possible net segments for a single net consisting of six pin clusters. The two leftmost cells on the second row up from the bottom give rise to a switchable net segment.

The first phase of the global router examines each net separately. The first operation in this phase is to generate the graph of possible pin cluster interconnections. In general, this graph contains switchable net segments as well as cycles. Next, the extra edges in the graph, that is, those edges creating cycles as shown in Figure 3.11, are eliminated by generating the minimum spanning tree (MST). The MST approach is sufficient since it has been observed that, on average, more than 98 percent of the nets have no extra edges. The final operation of the first phase is to generate two net segments for each switchable net segment.

The phase one algorithm, $GR_1()$, will now be presented in a pseudo-C language format. In the description of the algorithm, the function $r(P)$, where P is a specific pin cluster, returns the row number to which the cluster belongs. If the nets are numbered from 1 to N, then the function $GR_1(n)$ is called N times, where $n \in \{1, 2, \cdots, N\}$.

$GR_1(n)$ {

build_Q();

/* The pin clusters for the net are determined. Furthermore, the average x coordinate of the constituent-pin locations is ascertained for each pin cluster. Next, the pin clusters are sorted by their average x coordinate, and are inserted in a first-in, first-out queue Q. */

$G(n) = G(V, E) = G(V, \varnothing)$;

/* The initial edge list E for the graph $G(n)$ is empty. The node list V consists of the set of pin clusters. */

while($(P = \text{pop}(Q)) \neq$ NULL) {

/* Select leftmost unexamined pin cluster P by popping queue Q. */

U = nearest cluster to the right of P and one row above, that is, $r(U) = r(P) + 1$. Let U_x represent the x location of U, where $U_x = \infty$ if U is null.

M = nearest cluster to the right of P and on the same row: $r(M) = r(P)$. Let M_x represent the x location of M, where $M_x = \infty$ if M is null.

D = nearest cluster to the right of P and one row below: $r(D) = r(P) - 1$. Let D_x represent the x location of D, where $D_x = \infty$ if D is null.

if($D_x < M_x$) {
add_edge(P , D) ;

/* Add an edge connecting nodes P and D to the graph $G(n)$ */

}
if($U_x < M_x$) {
add_edge(P , U) ;
}
if($M_x \leq U_x$ && $M_x \leq D_x$) {
add_edge(P , M) ;
}

}

MST(n) ;

/* At this point, all of the possible edges connecting the nodes have been added to the graph for this particular net. Kruskal's algorithm is then applied to generate the MST.[1] The use of Kruskal's algorithm to generate the minimum spanning tree for a net does not necessarily yield the best possible global route for the net based on the total channel density metric. However, it does minimize the interconnect length. Nonetheless, a 300-cell circuit was examined to determine the number of *extra edges* in the graph $G(n)$. The 300-cell circuit had a total of 349 nets and only 7 of the nets had extra edges. Thus, Kruskal's algorithm only had an effect on 2 percent of the nets (for all of the other nets, the graph generated by steps 3 through 6 above was already a tree). The 7 extra-edge nets had an average of 53 segments and an average of 6 extra edges. With such a vast number of segments for the extra-edge nets, the use of Kruskal's algorithm appears to be a reasonable choice since these nets were observed to connect to positions spanning virtually the entire chip. That is, it isn't likely that another approach could do much better with nets connecting to over 50 positions all over the chip. */

implement_net_segments(n) ;

---

[1] J. Kruskal, "On the Shortest Spanning Subtree of a Graph and the Traveling Salesman Problem," (1956).

/* For each edge of the tree representing the global route of the specific net n, one pin from each cluster is selected to form the actual net segment. In the case of an edge connecting two clusters on the same row, it is determined if this is a switchable net segment. If the segment is switchable, then two pairs of pins are selected. One pair is for the segment routed in the channel above the row and another pair is for the segment routed in the channel below the row. */

}

3.4.3 The second phase of the global router

The purpose of the second phase of the global router is to select from among the choices presented by the switchable net segments in such a manner as to achieve an apparent local minimum in the total channel density, expressed by Eqn. 3.36. The switchable net segments for *all* of the nets are considered simultaneously. It is noteworthy that the aforementioned 300-cell circuit had a total of 1032 net segments, of which 389 were switchable net segments. This implies that a remarkable 38 percent of the net segments were switchable for this circuit. A very similar percentage was noted for each of more than 10 industrial circuits. Hence it is not surprising that the proper choices of channels for the switchable net segments yield substantial reductions in the total channel density.

The algorithm for the second phase is written in the form of simulated annealing. However, $T = 0$ is maintained throughout the execution of the algorithm. Thus, the second phase is essentially a random-interchange (of a switchable net segment) algorithm. The various aspects of the implementation of simulated annealing are now presented in detail.

1. The function generate()

The generation of new states takes on a very simple form for this implementation. A net segment n_k is randomly selected from a list of all switchable net segments. A new state is generated by switching the routing of the net segment n_k from its current channel to the opposite channel.

2. The cost function

The cost function is given by Eqn. 3.36, the total channel density. If the generation of a new state results in $\Delta C = 0$, then switching the channels of the switchable net segment has no effect on the total channel density. In an effort to aid the process of detailed routing, the switchable net segment is placed in the channel which has the least *congestion* along the span of the net segment. The degree of congestion, C_{ab}, for a partial span of a channel (from $x = a$ to b) is defined by the difference between the density of the channel and the maximum number of nets crossing a cut line (x) belonging to the partial span (that is, for $a \leq x \leq b$. The magnitude of C_{ab} is inversely related to the congestion over the span from a to b. That is, a value of C_{ab} at or near zero implies that the number of nets crossing a cut line in the span from a to b is at or near the density of the channel.

Note that the generation of a new state implies removing a segment from one channel and then placing it in another channel. The removal of the segment from the first channel can only decrease the total channel density by one or leave it the same. Similarly, the addition of a net segment to the second channel can only increase the total channel density by one or leave it the same. Hence, as a result of the generation of a new state, the cost function either increases in value by 1, decreases by 1, or remains the same.

The behavior of simulated annealing with such a cost function was investigated. Table 3.14 shows the results for a global routing problem in which generate() and the cost function were implemented as described above. In this case, the change in the value of the cost function can take on 3 possible values, namely $\Delta C \in \{ +1, -1, 0 \}$. The second column in the table has the average final total channel density for 20 trials in which a full range of values of T were used ($\alpha = 0.9$ and the initial T was such that almost all new states were accepted at this T) for the simulated annealing implementation. The third column has the average final total channel density for 20 trials in which $T = 0$ was maintained throughout the runs. Note that there was no observable difference in performance between the *annealed* and *non-annealed* cases. The same behavior was noted for several other circuits from Gould-AMI, with cell counts varying from 100 to 1500. Hence the decision to go with $T = 0$ insofar as the CPU time is smaller for this greedy method.

Trials	Average Final Total Density (annealed)	Average Final Total Density (non-annealed)
20	243.80	243.65

Table 3.14 Effect of simulated annealing for $\Delta C \in \{ +1, -1, 0 \}$.

3. The stopping criterion

Since $T = 0$ is maintained throughout the run, there was no implementation of an inner loop criterion. Let N_{sw} represent the total number of switchable net segments. Further, define A_{sw} as the number of attempted new states per switchable net segment. The *stopping criterion* is satisfied (1) if a number of new states equal to ($N_{sw}A_{sw}$) have been attempted, or (2) if a consecutive number of proposed new states equal to ($\mu N_{sw}A_{sw}$) have failed to gain acceptance.

Tests were conducted on industrial circuits, with cell counts ranging from 50 to 3000, to determine suitable choices for μ and A_{sw}. When $\mu \geq 0.3$, and when $A_{sw} \geq 30$, the final *total* channel density was always equal to or one above the best observed value for CPU times which more than 20 times greater than the $\mu = 0.3$ and $A_{sw} = 30$ case. (This was measured by trying $A_{sw} = 600$ and $\mu = 1.0$, and then reducing each of the two parameters until an impact on the total channel density could be observed.) When the global router has completed, the total channel density has converged within the constraints of the current cell placement and the implementation of the stopping criterion.

3.5 The Algorithm for Stage 3 of TimberWolfSC

Stage 3 is the placement, and global routing, refinement stage. After the global router has finished, a profile of the congestion of a typical channel resembles the plot shown in Figure 3.12.

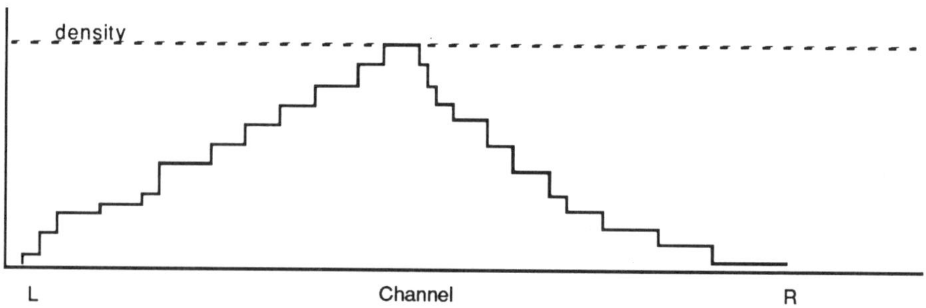

Figure 3.12 Typical plot of channel congestion versus channel position at the end of stage 2.

That is, the minimization of the total interconnect length often results in congestion peaks in the middle portions of the channels. The purpose of stage 3 is to attempt neighboring-cell interchanges in the maximum-congestion regions in an attempt to reduce the total channel density. That is, stage 3 seeks to minimize the total channel density by refining the standard cell placement. Note that at the end of stage 2 of TimberWolfSC, T has been reduced to nearly zero. Consequently, in stage 3, a greedy iterative improvement algorithm is used. This random-interchange algorithm is characterized by a generation of new configurations function and a cost function.

The function generate() has the following form: First, a row is randomly selected. Next, a cell i is randomly selected which lies in a *maximum-congestion* region for this row. Maximum-congestion regions for a given row are defined to be those portions of the span of the row in which the congestion in either the channel above the row or the channel below the row is at the maximum value (density) for the respective channel.

The remainder of generate() closely resembles that used in the stage 2 simulated annealing algorithm. Define L to be the left neighbor of cell i and R to be the right neighbor of cell i. If L is null, then the next new configuration attempted is an interchange of cells i and R. On the other hand, if R is null, then the next new configuration attempted is an interchange of cells i and L. If neither L nor R is null, one of them is randomly selected to be interchanged with i. If this new configuration is not accepted, then the next new configuration is generated by interchanging i with the other neighbor. Finally, the last step in a call to generate() results in an attempted orientation change for cell i. As in stage 2, the neighboring pairs of cells are interchanged in such a manner as to not create any overlap, nor to create any empty space between cells.

The cost function is again given by Eqn. 3.36, the total channel density. If the change in the total channel density is less than or equal to 0, then the new configuration is retained.

For each new configuration generated, the change in the value of the cost function is computed, with the following considerations: The nets connected to the cell or cells involved in the generation of the new configuration must be resubmitted to the global router. These nets must be subjected to the first phase of the global router. *All* of the nets of the circuit are then subjected to the second phase of the global router, where the optimum choice of channel is sought for each switchable net segment. Note that nets other than those subjected to the first phase must be resubjected to the second phase since some of them had their positions chosen based on the positions of the *moved* nets. After the conclusion of the second phase of the global router, the new value of the total channel density is compared to the previous value. If the new value is less than or equal to the previous, then the interchange of cells or the cell-orientation change is accepted.

The stopping criterion for stage 3 is satisfied when the total channel density has not been reduced for a successive number of trials equal to: $\lambda(N_c + N_f)$, where N_c is the number of standard cells and where N_f is the number of inserted feed-through cell instances. A suitable value of λ is a floating point number such that $0.5 \leq \lambda \leq 2.0$.

The effect of stage 3 is to even out the congestion in the channels, while reducing the peak congestion, as illustrated in the transition from Figure 3.12 to 3.13. For eight Gould-AMI circuits, ranging in size from 100 to 1500 cells, the TEIL increases during stage 3 were always under one percent.

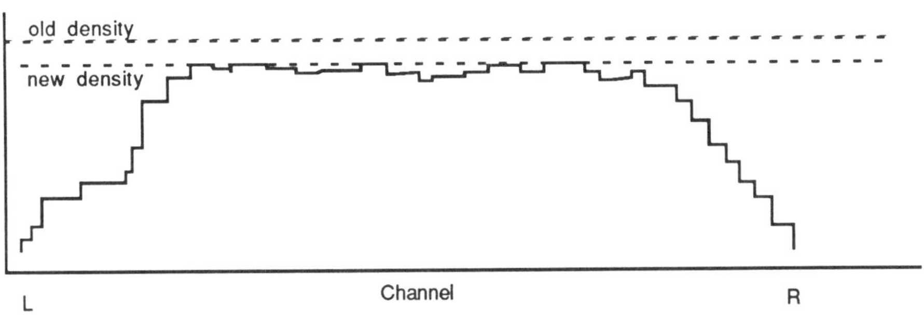

Figure 3.13 Typical plot of channel congestion versus position at the end of stage 3.

3.6 TimberWolfSC Results

3.6.1 Comparisons taken at the end of stage 1

The TimberWolfSC standard cell placement and global routing package was compared with several automatic and manual placement strategies, in which only stage 1 of TimberWolfSC was used. That is, the global router was not used, nor was the local congestion minimization (placement and global routing refinement) performed. The comparisons are summarized in Table 3.15. Each company used their own detailed router to implement the interconnections for both their placements and the TimberWolfSC placements.

Circuit	From	No. Cells	TEIL Red.	Area Reduction
A-F	AMI	2700	75%	70%
A-G	AMI	1500	–	40%
A-A1	AMI	1500	45%	30%
A-A2	AMI	1500	37%	25%
A-B	AMI	1000	57%	31%
I-1	Intel	1000	–	10%
I-2	Intel	1000	–	25%
H-1	Hughes	800	32%	6%

Table 3.15 Stage 1 comparisons with industrial methods.

Note that the TEIL was reduced by up to 75 percent, and that the chip area was reduced by up to 70 percent, in comparison to Gould-AMI's automatic placement.[1] [2] Circuit I-1 was placed manually by Intel, and the TimberWolfSC placement not only reduced the area by 10 percent, but 30 percent fewer feed-through cells were needed to complete the routing. The TimberWolfSC placement for circuit H-1 was also compared to a manual placement. The TEIL and the chip area were reduced, and furthermore, 26 percent fewer feed-through cells were required.

3.6.2 The effectiveness of the global router

The results given in Table 3.16 demonstrate the effectiveness of the global router. The column entitled *initial tracks* is the total channel density just prior to the execution of the global router in stage 2 of TimberWolfSC, in which a switchable net segment is arbitrarily routed in the channel above its row. The column entitled *final tracks* is the total channel density following the execution of the global router. Note that the total channel density was reduced by 7 to 16 percent for the six circuits of various sizes.

Circuit	No. Cells	Initial Tracks	Final Tracks	Track Red.
EX1	800	321	297	7%
EX2	2500	726	651	10%
EX3	469	134	114	15%
EX4	469	160	141	12%
EX5	300	125	105	16%
EX6	180	68	60	12%

Table 3.16 The effect of the global router.

Tests were also performed using both stages 1 and 2 of TimberWolfSC. The global router portion of TimberWolfSC was also interfaced to the CIPAR placement and routing package developed by Gould-AMI. For the largest circuit (2700-cell CktF), the

---

[1] C. Horng and M. Lie, "An Automatic/Interactive Layout Planning System for Arbitrarily-Sized Rectangular Building Blocks," (1981).

[2] M. Lie and C. Horng, "A Bus Router for IC Layout," (1982).

global router reduced the area by another 8 percent. This was in addition to the area savings achieved by the minimization of the TEIC. A total area savings of 75 percent was achieved for CktF when both the TimberWolfSC placement optimization and the global router (stages 1 and 2) were applied. The results are summarized in Table 3.17, showing the additional area reductions due to use of the global router and also the overall area reductions as a result of using both TimberWolfSC placement and global routing. Note that the CPU time for the global router is quite modest.

Circuit	No. Cells	Global Router Area Red.	Final Chip Area Red.	CPU Time (VAX 780 mins.)
CktF	2700	8%	75%	2.0
CktG	1500	8%	45%	1.0
CktA	1500	6%	34%	1.0
CktB	1000	6%	35%	0.5

Table 3.17 Global router results for Gould-AMI circuits.

3.6.3 The effectiveness of stage 3 of TimberWolfSC

The effectiveness of stage 3, the placement and global routing refinement stage, of TimberWolfSC is shown in Table 3.18 for five industrial circuits. The third column in Table 3.18 is the total channel density after the completion of stage 2, that is, after the global router has been applied to the placement achieved by the minimization of the TEIC. The column entitled *final tracks* is the total channel density following the conclusion of stage 3. Note that the total channel density reductions ranged from 4 to 14 percent. It is apparent from the table that the reduction in the total channel density as a result of stage 3 is quite substantial.

Circuit	No. Cells	Initial Tracks After Stage 2	Final Tracks	Track Red.
EX1	800	297	270	9%
EX2	2500	485	461	5%
EX3	469	114	109	4%
EX5	300	105	90	14%
EX6	180	60	54	10%

Table 3.18 The effect of stage 3 of TimberWolfSC.

3.6.4 TimberWolfSC comparisons including stage 3

The TimberWolfSC placement and global routing package was compared with another commercially developed placement program. Two double-metal technology circuits were selected for comparison. The first consisted of 2500 cells and 2 macro blocks. The second contained 469 cells. The results are shown in Table 3.19.

Circuit No. Cells	Placement Performed By	Inserted Feeds	Built-In Feeds	Total No. Feeds	Final Total Chan. Density
2500	other	226	1003	1229	652
2500	TW3.2	57	747	804	417
469	other	67	26	93	141
469	TW3.2	0	76	76	109

Table 3.19 Results of TimberWolfSC versus a commercial placement package.

Note that for the larger (2500-cell) circuit, TimberWolfSC produced a total channel density which is 36 percent smaller than that produced by the commercial placement package. For this circuit, TimberWolfSC required 35 percent fewer feed-through paths, including 75 percent fewer actual feed-through cells. Furthermore, the TimberWolfSC placement had a TEIL which was 37 percent smaller. TimberWolfSC required approximately 15 hours of CPU time on an IBM 3081K to achieve the results for the larger circuit shown in Table 3.19.

For the smaller (469-cell) circuit, TimberWolfSC produced a total channel density which is 23 percent smaller than that produced by the commercial placement package. For this circuit, TimberWolfSC required 18 percent fewer feed-through paths, including a 100 percent reduction in the actual feed-through cells required. Furthermore, the TimberWolfSC placement had a TEIL which was 17 percent smaller. TimberWolfSC used approximately one hour of CPU time on an IBM 3081K to achieve the results for the smaller circuit shown in Table 3.19.

TimberWolfSC was used to accomplish the placement of 13 blocks, each composed solely of rows of standard cells, on the Intel 80386 microprocessor chip. Each block required the placement of 100 to 350 cells. For each of the 13 blocks, TimberWolfSC was able to meet the strict hand layout projections for area as directed by the overall chip plan.[1]

The performance of TimberWolfSC was also benchmarked at Sony Corp. in Japan. The results for a 473-cell circuit are shown in Table 3.20. The first column contains the data for manual placement followed by automatic routing by their program called Sweet. The second column contains the results of the automatic layout performed by Sony's FLEX layout package. This package consists of the Scoop automatic placement program and the Sweet detailed router. The placement generated by TimberWolfSC was also routed by Sweet. Note that the TimberWolfSC result is substantially better than the manual placement, although only slightly better than the FLEX result for this small circuit.

[1] J. Krauskopf, Private Communication, (1986).

	Manual	FLEX	TimberWolfSC
Area	3.48 × 3.60	3.39 × 3.41	3.36 × 3.38
Normalized	100.0%	92.3%	90.7%

Table 3.20 Results of TimberWolfSC versus layouts performed by Sony.

In benchmark tests on a 2000-cell circuit performed at Texas Instruments, in comparison to two widely available commercial standard cell placement and routing systems, TimberWolfSC was the only package to achieve a layout with the minimum possible chip size.[1] This minimum was due to *pad limitation* which arose from the inability to further compact the pad cells along the periphery of the chip. Unfortunately, I was not able to gain the release of additional statistical data.

[1] M. Roberts, Private Communication, (1986).

Chapter 4

Macro/Custom Cell Chip-Planning, Placement, and Global Routing

4.1 Introduction

This chapter presents the algorithms and the implementation of a new macro/custom cell placement and global routing package. This package, named *TimberWolfMC*, makes extensive use of the simulated annealing algorithm presented in Chapter 2. TimberWolfMC is applicable to circuits containing cells of any rectilinear shape. Furthermore, the cells may have fixed geometry including pin locations (macro cells) or the cells may have an estimated area with a specified aspect ratio range, and with pins that need to be placed (custom cells). The pins for a custom cell may be specified in several possible ways: (1) a pin may be given a particular fixed location, (2) a pin may be assigned to a particular edge or edges of a cell, (3) a pin may belong to a group of pins which may be assigned to a particular edge or edges of a cell, or (4) a pin may belong to a group of pins which is assigned a particular sequential ordering as well as a particular edge or edges of a cell. The placement of a single pin, a group of pins, or a sequence of pins may be specified as being restricted to either one cell edge, two cell edges, or any of the edges.

TimberWolfMC also permits the custom cells to have aspect ratios in a continuous or discrete range. Furthermore, several alternative instances for a custom cell may be specified, whereby TimberWolfMC is to select the one which is most suitable. The selection of aspect ratio and/or instance is guided by the minimization of the Total Estimated Interconnect Cost (TEIC) and by the geometry of the empty space allotted for the cell as influenced by the neighboring cells. TimberWolfMC places circuits consisting entirely of macro cells as well as circuits consisting entirely of custom cells. Furthermore, the program will place circuits consisting of a combination of macro and custom cells. Consequently, TimberWolfMC is applicable to chip planning problems.

As presented in the previous chapter, simulated annealing has been an effective algorithm upon which to base a standard cell placement and global routing package.[1 2 3 4 5] The use of simulated annealing was therefore considered for the generalized macro/custom cell placement problem. Simulated annealing is used to minimize the TEIC. The TEIC calculation is based on the exact pin locations (as opposed to assuming that all of the pins associated with a cell have a location equal to the center of the cell). Consequently, all eight possible orientations are considered for each cell. A new algorithm for accurately estimating the interconnect area around the individual cells, introduced in Chapter 6, is used in TimberWolfMC. This algorithm *dynamically* estimates the interconnect area in the sense that after each new placement configuration, the interconnect-area estimate is updated. This interconnect-area estimation algorithm has resulted in the generation of placements which require virtually no placement modification during or after detailed routing.

The placement algorithm proceeds in two distinct stages. During the first stage, the interconnect area around the individual cells is determined using the dynamic interconnect-area estimator. That is, as each cell is moved, its effective area is adjusted. A simulated annealing algorithm is used to minimize the TEIC. The second stage of TimberWolfMC consists of several executions of a placement-refinement algorithm. Each execution consists of three steps: (1) a channel definition step, (2) a global routing step, and (3) a placement-refinement step. The information obtained in step 2 is used to compute the density of all of the channels, which then permits accurate interconnect-area determination. The placement of the cells is then refined in step 3 to reflect the required interconnect area. A low-temperature simulated annealing algorithm is used to accomplish step 3. Three such placement-refinement steps are sufficient for the final Total Estimated Interconnect Length (TEIL) and the final chip area to converge.

Total interconnect length savings of 8 to 49 percent were achieved in experiments on 9 industrial circuits. Furthermore, for the 9 examples, circuit-area reductions ranged from 4 to 56 percent versus a variety of other placement algorithms.

The remainder of this chapter contains five sections. Section 4.2 presents some details of the general TimberWolfMC methodology. In Section 4.3, the algorithms for the first stage of TimberWolfMC are presented. The algorithms for the second stage, including the placement-refinement procedure, are described in Section 4.4. The experimental results are presented in Section 4.5. Finally, the conclusion is the subject of Section 4.6.

---

[1] C. Sechen and A. Sangiovanni-Vincentelli, "The TimberWolf Placement and Routing Package, (1984).

[2] C. Sechen and A. Sangiovanni-Vincentelli, "The TimberWolf Placement and Routing Package," (1985).

[3] C. Sechen and A. Sangiovanni-Vincentelli, "TimberWolf3.2: A New Standard Cell Placement and Global Routing Package," (1986).

[4] C. Sechen and K. W. Lee, "An Improved Simulated Annealing Algorithm for Row-Based Placement," (1987).

[5] C. Sechen, D. Braun, and A. Sangiovanni-Vincentelli, "ThunderBird: A Complete Standard Cell Layout Package," (1988).

4.2 The General TimberWolfMC Methodology

This section focuses on three aspects of TimberWolfMC which provide flexibility and robustness to the general methodology and algorithms. The first subsection presents the algorithms for handling rectilinear cells. The next subsection describes the generation of the initial placement configuration. Finally, the last subsection introduces the custom-cell pin placement methodology.

4.2.1 Algorithms for handling rectilinear cells

TimberWolfMC handles cells of any *rectilinear* shape. That is, each vertex of a cell is an intersection point of two edges meeting at right angles. In order to facilitate the conversion of a general rectilinear cell to a union of nonintersecting rectangular *tiles* exactly covering the cell, the vertex list is given in a clockwise order starting from the leftmost of the lowest vertices (the lower left corner in the case of a rectangular cell). During both stages of TimberWolfMC it is very often necessary to calculate the amount of overlapping between a pair of cells. This is implemented by first noting whether or not there is any overlapping between the bounding boxes (the smallest enclosing rectangles) of the two cells. If there isn't, then there can be no overlap between the two rectilinear cells. On the other hand, if the bounding boxes overlap, then the area of overlap (if any) is computed by summing the overlaps over all pairs of tiles, one from each cell. This subsection first illustrates the details of the algorithm referred to as bust(), which is used to break down the rectilinear cells into a union of nonintersecting rectangular tiles fully covering the cell.

At the conclusion of the TimberWolfMC placement, it is necessary to convert the tiles for each cell back to a vertex list prior to generating the output (that is, the placement information). It is not sufficient to store the original vertex list since a given cell may have a new shape as well as a new orientation. Consequently, this section concludes with a description of the algorithm unbust() which regenerates a vertex list from a set of rectangular tiles. The unbust() algorithm assumes that the list of tiles was created by the bust() algorithm.

4.2.1.1 The bust() algorithm

The bust() algorithm breaks down the rectilinear shape into rectangular tiles of maximum possible width, starting from the bottom of the cell. The following steps outline the bust() algorithm, which is applied to each rectilinear cell C_i.

```
bust() {
        {A} = V_l(C_i) ;
                /* The function V_l returns the list of all vertices of the rectilinear cell given in its argu-
                ment field, which then forms the set {A}. */
        while( {A} ≠ ∅ ) {
                P_k = the leftmost of the lowest vertices in {A} ;
                P_l = the next leftmost of the lowest vertices in {A} ;
                P_m = leftmost of next lowest vertices in {A}, where P_kx ≤ P_mx ≤ P_lx ;
                        /* The x coordinate of vertex P_m is designated as P_mx. */
                output = {R} = BB(P_k, P_l, P_m) ;
                        /* Output the bounding box of the points P_k, P_l, and P_m. Let {R} be the set
                        of vertices of the bounding box, where R_1, R_2, R_3, and R_4 represent the four
                        vertices, enumerated in a clockwise fashion from the lower left corner. */
                {A} = {A} - {R} ;
                        /* Delete from {A} those vertices of {R} which are in {A}. */
                {A} = {A} + {{R} - {A}} ;
                        /* Add to {A} those vertices of {R} which are not in {A}. Note that {R_1} and
                        {R_4} are necessarily in {A}, and hence always have to be removed from {A}. */
        }
}
```

The bounding boxes generated by the function BB() in bust() constitute the set of rectangular tiles exactly covering the rectilinear cell. As an example of the operation of bust(), consider the sequence of events shown in Figures 4.1a-e. Note that five tiles (T_1, T_2, T_3, T_4, and T_5) are created which exactly cover the shape of cell C_4. The vertices P_k, P_l, and P_m used to generate each tile are indicated in each of the figures.

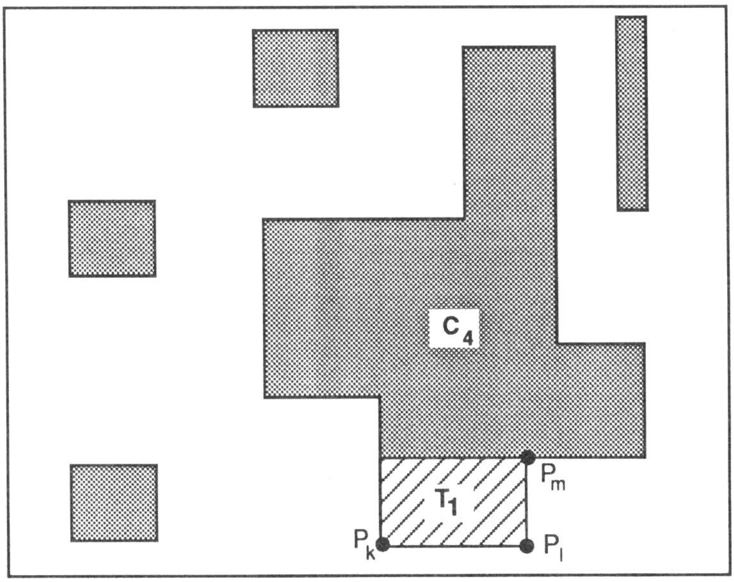

Figure 4.1a Five tiles (T_1, T_2, T_3, T_4, and T_5) are created which exactly cover the shape of cell C_4. The vertices P_k, P_l, and P_m used to generate each tile are indicated in the figure.

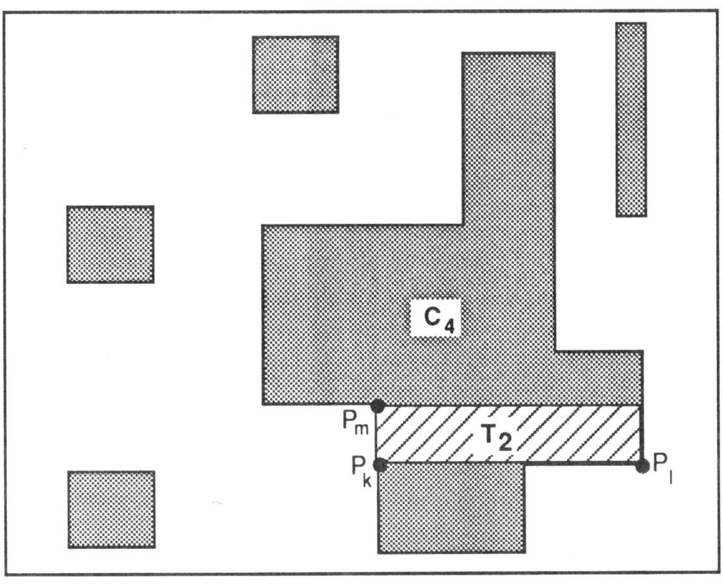

Figure 4.1b The generation of tile T_2.

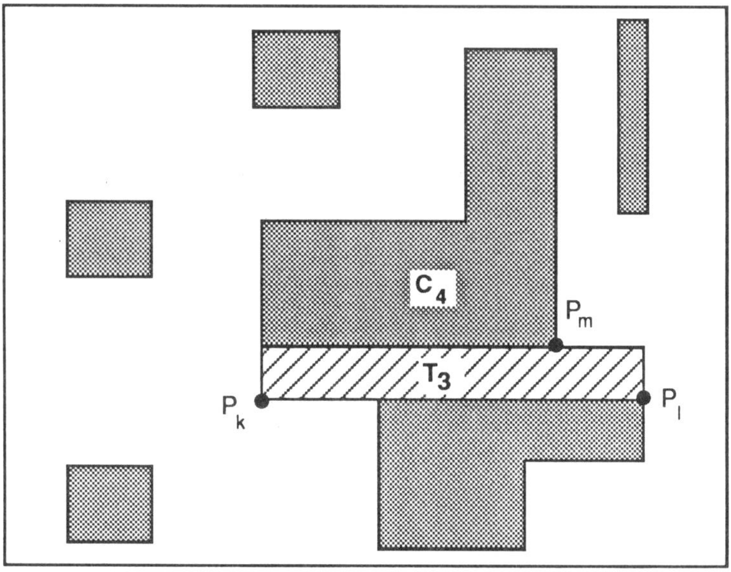

Figure 4.1c The generation of tile T_3.

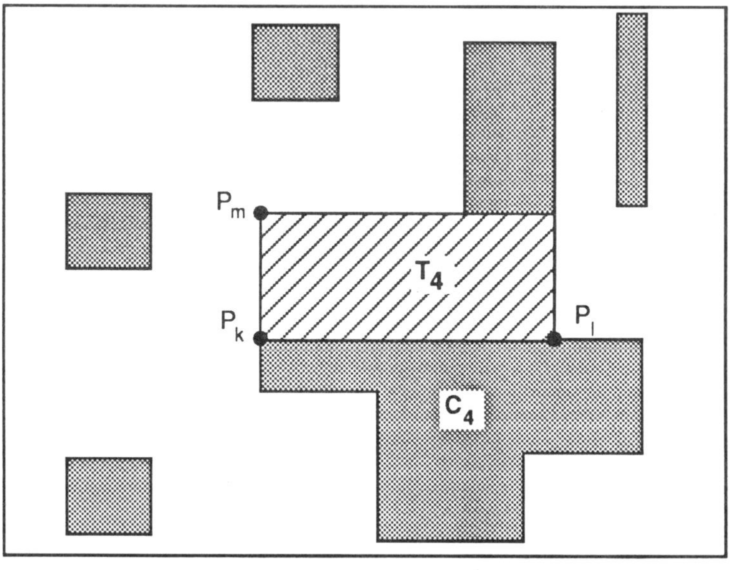

Figure 4.1d The generation of tile T_4.

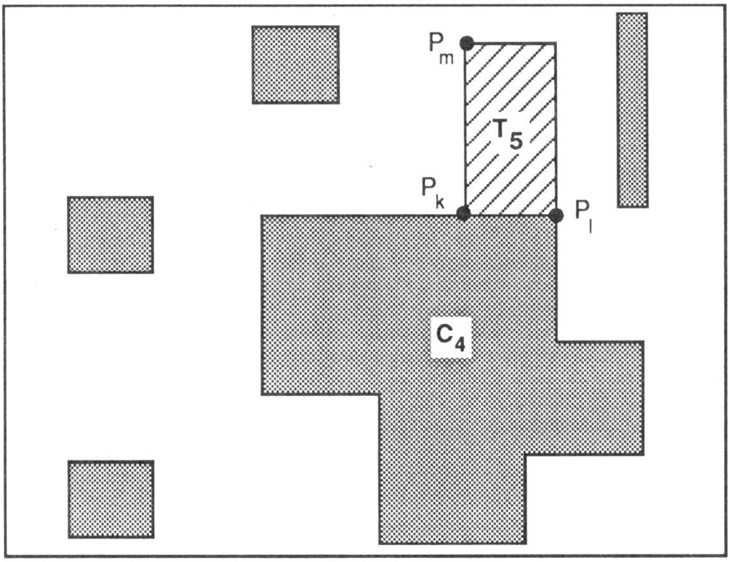

Figure 4.1e The generation of tile T_5.

4.2.1.2 The unbust() algorithm

The purpose of the unbust() algorithm is to output a list of vertices in the clockwise direction, starting from the leftmost of the lowest vertices, defining a rectilinear shaped cell where the input is a set of rectangular tiles exactly covering the rectilinear cell. The implementation of unbust() is less generally applicable than the function bust(). For example, bust() will generate the set of rectangular tiles covering the rectilinear shape for *any* such rectilinear shape. However, unbust() will only generate a proper vertex list from the rectangular tiles if the tiles were created by bust(). Furthermore, unusual rectilinear shaped cells (for example, a rectilinear doughnut-shaped cell), will not be properly handled by unbust(). However, for any cell which does not fully enclose an empty region, the function unbust() will generate the proper vertex list defining the rectilinear shaped cell (again assuming that the tiles were generated by bust()).

The overview of the unbust() algorithm is as follows:

(1) Accumulate the set of vertices $\{A\}$ (referred to as the *points* below) of the various tiles describing the rectilinear shape, four vertices per tile. For example, the set of points obtained from the five tiles generated from cell C_4 in Figure 4.1a is shown in Figure 4.2a.

(2) Eliminate the coincidental points from $\{A\}$. Each encircled point in Figure 4.2a actually represents a coincident pair of points.

(3) The leftmost of the lowest vertices in $\{A\}$ is necessarily part of the vertex list for the description of the rectilinear shape. Hence, starting with this vertex, labeled P_0 in Figure 4.2a, output the points of $\{A\}$ in a clockwise fashion in which the cell area always lies to

the *right* of a generated point-to-point segment. More specifically, if the last two points generated (P_1 and P_2, in that order) are such that: (a) If P_2 is directly above P_1, then the search for the next point P_3 begins first in the left direction. If no such point can be found, then the next search direction is directly upward. Finally, if no such point was found, then the next search direction is directly to the right. (b) If P_2 is directly to the left of P_1, then the search for the next point, P_3, begins first in the straight downward direction. If no such point can be found, then the next search direction is directly left. Finally, if no such point was found, then the next search direction is directly upward. (c) If P_2 is directly to the right of P_1, then the search for the next point (P_3) begins first in the straight upward direction. If no such point can be found, then the next search direction is directly to the right. Finally, if no such point was found, then the next search direction is directly downward. (d) If P_2 is directly below P_1, then the search for the next point (P_3) begins first in the straight right direction. If no such point can be found, then the next search direction is directly downward. Finally, if no such point was found, then the next search direction is directly to the left.

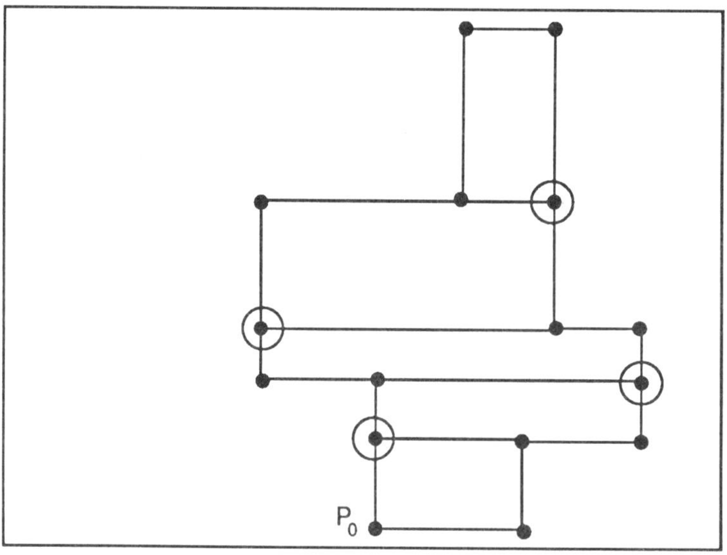

Figure 4.2a The set of points obtained from the five tiles generated from cell C_4 in Figure 4.1a. Each encircled point actually represents a coincident pair of points.

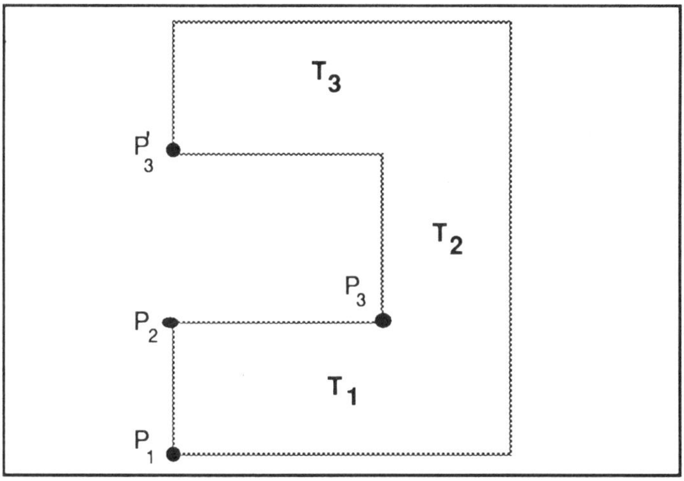

Figure 4.3 P_1 and P_2 indicate the previously generated pair of points. The search from P_2 finds point P'_3. However, P_2 and P'_3 do not belong to the same tile. The search continues from P_2 and ultimately finds point P_3. Points P_2 and P_3 are contained by tile T_1.

There is another restriction placed on the generation of points from P_2 above. Each sequential pair of points must lie on one of the boundaries of some particular tile. That is, if a single tile cannot be found such that a boundary of the tile contains both points, then this pair of points cannot constitute a part of a cell edge. As an example, in Figure 4.3, points P_1 and P_2 indicate the previously generated pair of points. From point P_2, the search to the left failed to find any points. Next, the upward search finds point P'_3. However, point P_2 is contained only by tile T_1 and point P'_3 is contained only by tile T_3. Hence, the pair of points P_2 and P'_3 do not constitute part of a cell edge. The search then continues from P_2 which ultimately finds point P_3, and furthermore, both points P_2 and P_3 are contained by tile T_1.

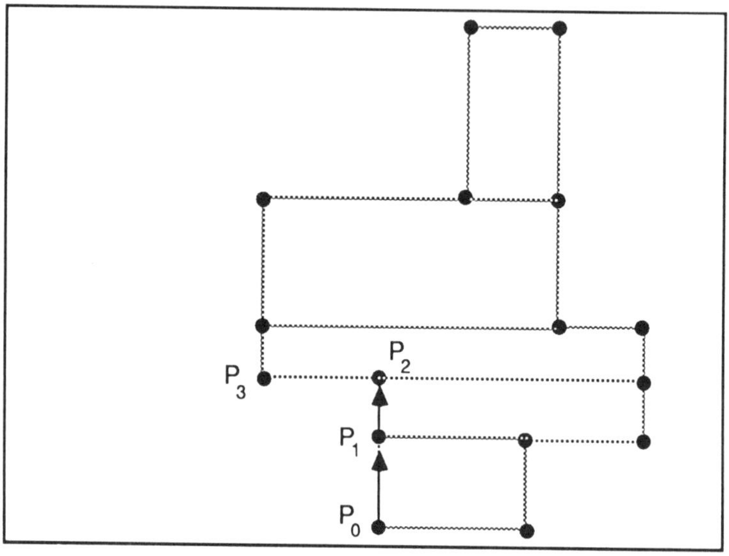

Figure 4.2b Segments from P_0 to P_1 and from P_1 to P_2 have already been generated. P_2 lies directly above P_1, implying that the first search direction is to the left. In this case, a point P_3 is found in the left direction.

In Figure 4.2b, segments from P_0 to P_1 and from P_1 to P_2 have already been generated according to the above methodology. At this point, P_2 lies directly above P_1, implying that the first search direction is to the left. In this case, a point P_3 is found in the left direction. Hence, the next segment produced is from P_2 to P_3 in Figure 4.2b. At the next step, as shown in Figure 4.2c, P_2 lies to the left of P_1. The first search direction is downward from P_2. Since this search fails to find a point, the next search direction is left of P_2. This search also fails to find a point. Finally, point P_3 is found on the upward search from P_2.

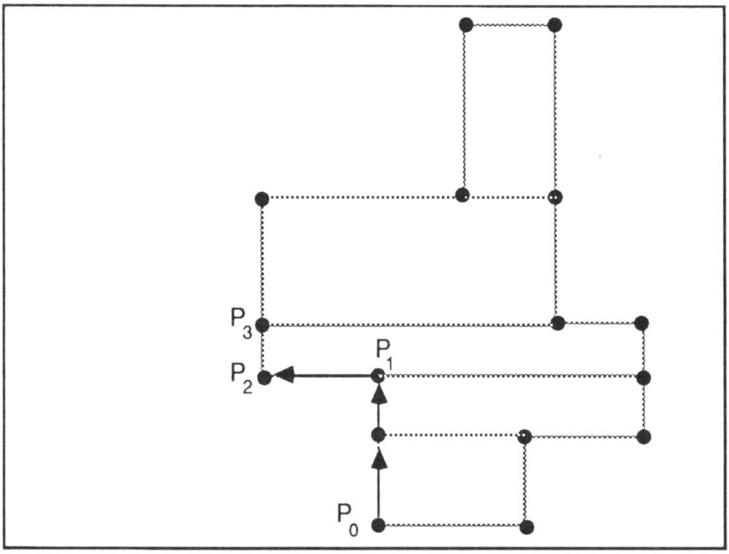

Figure 4.2c P_2 lies to the left of P_1. The first search direction is downward from P_2. This search fails to find a point. The next search direction is left of P_2. This search also fails to find a point. Finally, point P_3 is found on the upward search from P_2.

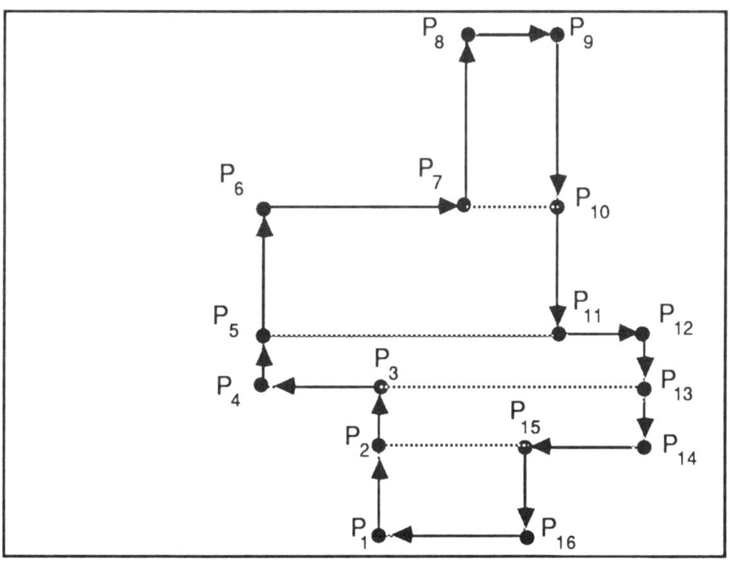

Figure 4.2d The sequence of 16 vertices defining the shape of the cell are illustrated.

(4) Figure 4.2d illustrates the sequence of 16 vertices defining the shape of the cell. Redundant points are removed next. If any sequence of three points generated in the manner described above are collinear, eliminate the middle (or 2nd) of the three points. The application of this rule results in the vertex sequence shown in Figure 4.2e. Note that vertices P_2, P_5, P_{10}, and P_{13} have been eliminated, and hence that 12 vertices are required to describe the shape of the rectilinear cell.

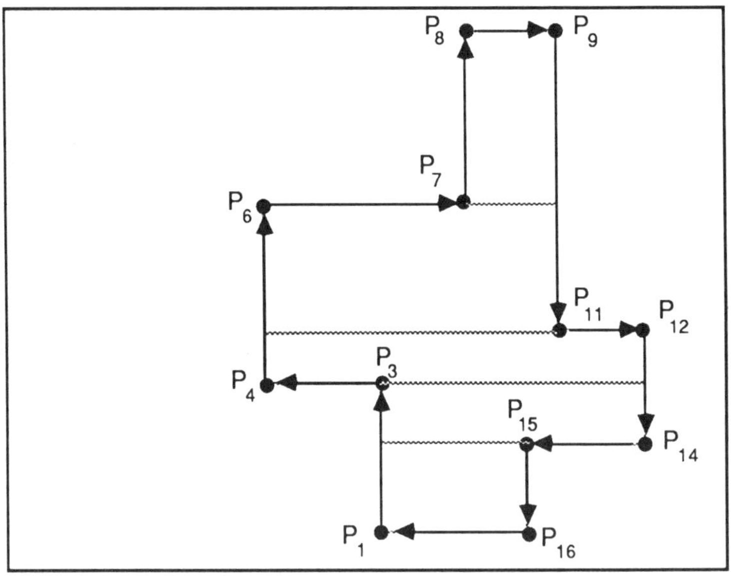

Figure 4.2e Redundant points (P_2, P_5, P_{10}, and P_{13}) have been removed.

A more formal description of the unbust() algorithm will now be presented. In the following, let P_1 and P_2 represent the most recently generated points for the vertex list of the rectilinear shape. The following four functions are defined as follows:

```
u() {
        /* u() returns the lowest point P3 remaining in the vertex list {A} which is directly above P2
        (that is, with the same x coordinate) and where the same tile contains both points P2 and P3. If no
        such point P3 exists, the function returns 0.   */
}

l() {
        /* l() returns the rightmost point P3 in the list {A} which is directly left of P2 (that is, with the
        same y coordinate) and where the same tile contains both points P2 and P3. If no such point P3
        exists, the function returns 0.   */
}

b() {
        /* b() returns the highest point P3 in the list {A} which is directly below P2 (that is, with the
        same x coordinate) and where the same tile contains both points P2 and P3. If no such point P3
        exists, the function returns 0.   */
}

r() {
        /* r() returns the leftmost point P3 in the list {A} which is directly to the right of P2 (that is,
        with the same y coordinate) and where the same tile contains both points P2 and P3. If no such
        point P3 exists, the function returns 0.   */
}
```

The variable *status* reflects the direction taken by the last point-to-point (P_1 to P_2) step. The following code is used: directly upward is represented by U, directly to the left is represented by L, directly to the right is represented by R, and directly downward is represented by D.

The unbust() algorithm is now presented:

```
unbust(C) {
        {A} = V(C) ;
           /* The function V returns the set of vertices for all tiles representing the shape of cell C. */
        {A} = elim( {A} ) ;
              /* Eliminate the coincidental points from the set {A}. */
        P2 = LL( {A} ) ;
        output( P2 ) ;
        {A} = {A} - P2 ;
        status = U ;
           /* LL() returns the leftmost of the lowest of the points in the set {A}. Output P2 as a vertex
           of the description of the rectilinear shape and eliminate P2 from the set {A}. Assign a value of
           U to the status variable, indicating that the point-to-point direction was upward. */
        switch( status ) {
                case U:
                        if( ( P3 = l() ) ≠ 0 ) {
                                status = L;
                                output( P3 );
                                {A} = {A} - P3;
                                P1 = P2 ;
                                P2 = P3 ;
                                continue ;
                        } else if( ( P3 = u() ) ≠ 0 ) {
                                status = U;
                                output( P3 );
                                {A} = {A} - P3;
                                P1 = P2 ;
                                P2 = P3 ;
                                continue ;
                        } else if( ( P3 = r() ) ≠ 0 ) {
                                status = R;
                                output( P3 );
                                {A} = {A} - P3;
                                P1 = P2;
                                P2 = P3;
                                continue ;
                        } else {
                                    /* There is no possible search direction. Thus the list of vertices
                                    describing the rectilinear shape of the cell is complete. */
                                elim_collinear() ;
                                    /* If there is any sequence of 3 collinear vertices (in the generated
                                    list of vertices), the middle vertex is eliminated. */
                                break ;  /* The algorithm has completed its course for this cell. */
                        }
```

```
case L:
        if( ( P3 = b() ) ≠ 0 ) {
                status = D ;
                output( P3 ) ;
                {A} = {A} - P3 ;
                P1 = P2 ;
                P2 = P3 ;
                continue ;
        } else if( ( P3 = l() ) ≠ 0 ) {
                status = L ;
                output( P3 ) ;
                {A} = {A} - P3 ;
                P1 = P2 ;
                P2 = P3 ;
                continue ;
        } else if( ( P3 = u() ) ≠ 0 ) {
                status = U ;
                output( P3 ) ;
                {A} = {A} - P3 ;
                P1 = P2;
                P2 = P3;
                continue ;
        } else {
                        /*  There is no possible search direction. Thus the list of
                        vertices describing the rectilinear shape of the cell is complete.
                        */
                elim_collinear() ;
                break ;
                        /*  The algorithm has completed its course for this cell.  */
        }
```

```
case R:
    if( ( P3 = u() ) ≠ 0 ) {
        status = U ;
        output( P3 ) ;
        {A} = {A} - P3 ;
        P1 = P2 ;
        P2 = P3 ;
        continue ;
    } else if( ( P3 = r() ) ≠ 0 ) {
        status = R ;
        output( P3 ) ;
        {A} = {A} - P3 ;
        P1 = P2 ;
        P2 = P3 ;
        continue ;
    } else if( ( P3 = b() ) ≠ 0 ) {
        status = D ;
        output( P3 ) ;
        {A} = {A} - P3 ;
        P1 = P2 ;
        P2 = P3 ;
        continue ;
    } else {
        elim_collinear() ;
        break ;
    }
```

```
        case D:
                if( ( P3 = r() ) ≠ 0 ) {
                        status = R ;
                        output( P3 ) ;
                        {A} = {A} − P3 ;
                        P1 = P2 ;
                        P2 = P3 ;
                        continue ;
                } else if( ( P3 = b() ) ≠ 0 ) {
                        status = D ;
                        output( P3 ) ;
                        {A} = {A} − P3 ;
                        P1 = P2 ;
                        P2 = P3 ;
                        continue ;
                } else if( ( P3 = l() ) ≠ 0 ) {
                        status = L ;
                        output( P3 ) ;
                        {A} = {A} − P3 ;
                        P1 = P2 ;
                        P2 = P3 ;
                        continue ;
                } else {
                        elim_collinear() ;
                        break ;
                }
        }
}
```

4.2.2 Generating the initial placement configuration

There are two basic steps which need to be completed in order to generate the initial placement configuration. The first is the computation of the core area and the second is the pad placement (around the core area). Experience gained from industry indicated that, in almost every case, the relative placement of the pads was predetermined by the circuit designers. Consequently, in TimberWolfMC, pads are placed in accordance with net list specifications, and are not changed by the simulated annealing algorithm.

The above considerations make it imperative that the exact positioning of the core with respect to the pads must be done by TimberWolfMC. That is, the net list data is used only to specify *general* localizations as to desired placement. For example, a group of pads may be given a specified *order* and may be designated to lie on the *top* side of the chip. In this fashion, TimberWolfMC can ensure that the initial placement configuration of the chip is such that the simulated annealing algorithm will be best able to find the placement of the highest possible quality every time.

TimberWolfMC gives the user the option of either specifying the target dimensions of the core, or supplying the desired aspect ratio of the core. In the latter case, the determination of the desired (or target) core area is made somewhat complex by the fact that the amount of interconnect area needed cannot be determined prior to the completion of the simulated annealing placement algorithm. During the course of the simulated annealing algorithm, the cells will be discouraged from extending beyond the core boundaries. If the target core area is excessive, the final aspect ratio generated by the placement algorithm may be quite far from the desired value. On the other hand, if the target core area is too small, either insufficient interconnect space will have been allotted between some of the cells, or cases of cell overlapping will be present.

TimberWolfMC uses the interconnect-area estimation algorithm presented in Chapter 6 to determine the expected interconnect area required for the chip. The sum of the actual cell areas and the expected interconnect area constitutes the value of A_t', the expected core area. If the user-supplied target aspect ratio is represented by a_r, then the target core width (W) is given by:

$$W = \frac{\sqrt{A_t'}}{\sqrt{a_r}} \tag{4.1}$$

Furthermore, the target core height (H) is given by:

$$H = \sqrt{a_r}\sqrt{A_t'} \tag{4.2}$$

4.2.3 Custom-cell pin placement

This subsection presents the details on pin placement with regard to custom cells (pins on macro cells are fixed in location). Pins on custom cells may be specified in several possible ways: (1) a pin may be given a particular fixed location, (2) a pin may be assigned to a particular edge or edges of a cell, (3) a pin may belong to a group of pins which may be assigned to a particular edge or edges of a cell, or (4) a pin may belong to a group of pins which is assigned a particular sequence ordering as well as a particular edge or edges of a cell. The placement of a single pin, a group of pins, or a sequence of pins may be specified as being restricted to either one cell edge, two cell edges, or any of the edges.

4.2.3.1 Introduction to the TimberWolfMC pin site methodology

Changes to the placement of the uncommitted pins, that is, those specified according to (2), (3), and (4) above, occur throughout the course of the simulated annealing algorithm in TimberWolfMC. On the order of 10^7 cell-orientation changes are performed during the course of a run for a circuit of over 100 cells. The time required to generate this many orientation changes would be a noticeable portion of the overall TimberWolfMC CPU time. As a consequence, all of the eight possible orientations for each cell are generated and stored at the outset. The geometry (including pin locations) for an orientation different than the current one can then be found by simply changing an array index.

The implication for custom cells is that all possible pin locations on the edges of the cell would have to be stored for each of the eight possible orientations. However, the possible pin locations on a custom cell can easily number into the thousands. In general, the amount of storage needed would be excessive. Furthermore, during the earlier portion of the simulated annealing run, that is, when T is relatively high, approximating the location of the pins is sufficient.

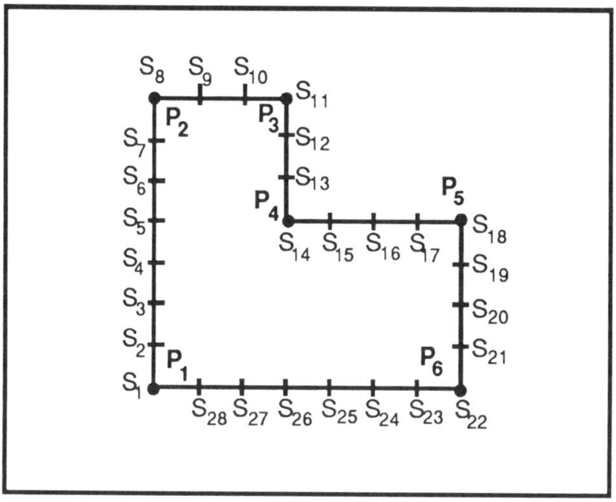

Figure 4.4 The pin sites are illustrated for the case of a custom cell containing six edges. Seven pins sites, labeled S_1 through S_7, are used to localize the locations of the pins currently placed on the first edge.

For these reasons, a strategy is used in which a specific number (bounded by an upper limit) of pin *sites* are defined for each edge of a custom cell. The pin sites are approximately evenly spaced along each edge. Figure 4.4 illustrates the case of a custom cell containing six edges. Seven pins sites, labeled S_1 through S_7, are used to localize the locations of the pins currently placed on the first edge. Similarly, pin sites labeled S_{18} through S_{21} are used to localize the locations of the pins currently placed on the fifth edge. Even though the locations of the sites are stored for each possible cell orientation, the amount of storage required is modest due to the upper bound on the number of pin sites per edge.

Each pin site is assigned a *capacity* in accordance with the number of pin locations encompassed by the site. As an example, consider a cell edge which has a length of 9200 units. If the design rules for the custom cells are such that a minimum center-to-center pin spacing is 5 units, then 1840 possible pin locations exist for this side. If 50 pin sites are permitted for this edge, a total of 40 of them will encompass 37 possible pin locations and 10 will encompass 36 possible pin locations. During the course of the simulated annealing algorithm, a penalty function approach is used to discourage more than 37 pins from being assigned to the first 40 of the sites and to discourage more than 36 pins from being assigned to the remaining 10 sites.

As another example, consider a cell edge which has a length of 1,000 units. If the design rules for the custom cells are such that a minimum pin-to-pin spacing is 50 units, 20 possible pin locations exist for this side. If up to 50 pin sites per edge are permitted, only 20 sites need be defined, each of which encompasses only one possible pin location. A capacity of one pin is then assigned to each of the sites, as is an associated length of 50 units.

The computation of the number of pin sites and the capacity for each site is further modulated by the specification of the aspect ratio range for a given cell. Consider an example of a vertical edge of height 9200 units, in which the current aspect ratio for the cell is 1.0. Suppose that the lower bound of the cell aspect ratio is 0.25. If, in the final placement, the cell aspect ratio is 0.25, the height of this edge will be only 4600 units. As a consequence, the number of pin sites for a cell edge and the capacities for the pin sites are computed for the *worst-case* aspect ratio for the cell. That is, if the edge is a *vertical* edge, then the worst-case aspect ratio is the lower bound for the aspect ratio of a cell. On the other hand, for horizontal edges, the worst-case aspect ratio is the upper bound for the aspect ratio of a cell. The worst-case analysis appears to restrict excessively the number of pins that are permitted to be assigned to a particular edge, particularly for a cell which is permitted a wide range of aspect ratios. However, analyses of industrial circuits have indicated that the number of possible pin locations on a custom cell tends to exceed greatly the number of pins to be placed on the cell.

Following the completion of the simulated annealing placement algorithm (in both stages), the uncommitted pins are assigned to specific locations within the span of the pin site to which they belong. In the following procedure, a single uncommitted pin is treated as a group (or sequence) having just one pin. The steps below are applied, in order, to each pin site s:

(1) If a portion of a group or sequence of pins fully occupies s, or if a group or sequence has pins which extend to each of the two neighboring pin sites, then the portion of the group or sequence belonging to s is the only source of pins in this site. The actual locations of the pins are then distributed as uniformly as possible within s. Note that a specific ordering of the pins within a group or sequence is always maintained within the pin sites.

(2) If an unplaced group or sequence begins or ends (but not both) within s, the pins in the group or sequence are assigned to the locations nearest to the relevant neighboring pin site.

(3) Unplaced groups or sequences (if any) which are contained entirely by s are assigned locations next. Suppose there are n_G such groups and/or sequences, represented by the set $\{G\}$. The *ideal* location for each pin $p \in \{G\}$ is then computed. The ideal location is that *available* position within s which is closest to the bounding box of the net to which p belongs, where the bounding box is computed without considering p. The ideal locations for each of the pins within a particular $g \in \{G\}$ are then averaged. This determines the ordering of the n_G groups within s. Also, for each group (which is not a sequence), the computed ideal locations determine the ordering of the pins within the group. The n_G groups

and/or sequences are then placed serially within p according to the procedure which follows.

> While there remains a group or sequence g to be placed in s:
>
> For each pin p in one of the groups or sequences remaining to be placed, determine whether or not it can be assigned to its ideal location subject to the constraint that sufficient locations are available for the unplaced pins which must lie to the left of p and to the right of p. That group or sequence which is able to obtain the most ideal locations is placed next. After a group or sequence has been placed, the ideal locations for the remaining pins are recomputed since there are now fewer locations available in s.

4.3 The Algorithm for Stage 1 of TimberWolfMC

This section presents the algorithmic details for the first stage of TimberWolfMC. Stage 1 uses a simulated annealing algorithm which seeks to minimize the TEIC while placing the macro/custom cells such that sufficient interconnect area is allotted between the cells. The value of the TEIC is determined by the sum over all nets of: (1) the horizontal span of the net times the horizontal weighting factor for the net, and (2) the vertical span of the net times the vertical weighting factor for the net. If all of the net-weighting factors have a value of 1.0, the TEIL is identically equal to the TEIC. During this first stage, the interconnect area around the individual cells is determined using the dynamic interconnect-area estimator described in Chapter 6. That is, as each cell is moved during the course of the simulated annealing algorithm, its effective area is adjusted.

4.3.1 The cost function

The cost function for the simulated annealing algorithm of stage 1 consists of three independent terms.

4.3.1.1 The first term in the cost function

The first term is a function $C_1()$ returning the TEIC. The stage 1 algorithm seeks to minimize this term while driving the next two terms of the cost function toward zero. The following terminology is used: (1) N_n is the total number of nets, (2) $X_s(n)$ returns the span of net n in the x, or horizontal direction, (3) $Y_s(n)$ returns the span of net n in the y, or vertical direction, (4) $H_w(n)$ returns the weighting factor for net n for the horizontal direction, and (5) $V_w(n)$ returns the weighting factor for net n for the vertical direction. The function $C_1()$ is then given by:

$$C_1() = \sum_{n=1}^{N_n} \{ X_s(n) H_w(n) + Y_s(n) V_w(n) \} \tag{4.3}$$

4.3.1.2 The second term in the cost function

The second term is the overlap penalty function, $C_2()$. The value returned by this function is given by:

$$C_2() = p_2 \sum_{i<j} O_c(i,j) \tag{4.4}$$

where p_2 is a normalization constants. The function $O_c(i,j)$ returns the total amount of overlap area between cells i and j. Formally, the function $O_c(i,j)$ is given by:

$$O_c(i,j) = \sum_{t_i=1}^{t_n^i} \sum_{t_j=1}^{t_n^j} O_t(t_i, t_j) \tag{4.5}$$

where the tiles comprising the area of the rectilinear cell i are represented by $t_i \in \{1, 2, \cdots, t_n^i\}$ and where the tiles comprising the area of cell j are represented by $t_j \in \{1, 2, \cdots, t_n^j\}$. Furthermore, the function $O_t()$ returns the common area between the two tiles specified as its arguments.

Sufficient interconnect space between the cells is maintained by, in effect, appending a border around the contours of each cell. The area occupied by each rectilinear cell is represented as a set of one or more nonoverlapping rectangular tiles. The edges of each tile for a cell are expanded outward by an amount determined by the interconnect-area estimation algorithm presented in Chapter 6.

The respective outward expansions of the tile edges are updated each time a cell participates in the generation of a new state in the simulated annealing approach. Hence the estimation of the interconnect area attributable to a given edge of a cell (or tile) is a *dynamic* quantity, depending on the location of the edge at any given time. It follows that if a cell is moved from a corner of the core area towards the center of the core area, then the effective area of the cell will increase. Similarly, if a cell is moved from the center towards a corner of the core area, then the effective area of the cell will decrease.

This penalty function was incorporated because of the usual difference in size and shape of the macro/custom cells. In the next subsection it will be seen that quite often two cells are selected for interchange which differ in size and/or shape. An exchange of location of these two cells often results in some overlap with one or more of the other cells. Furthermore, a single cell is often selected for displacement to a random new location. Since there is no prior check for the existence of one or more cells in the vicinity of this new location, some overlapping may well result.

The alternative to the overlap penalty function approach is, of course, to not permit overlapping. For example, when inserting a cell into an area, if insufficient space is available, then the other cells in the vicinity must be shifted away as necessary to accommodate the additional cell. This has the obvious disadvantage of destroying the local placement optimization process between the shifted and unshifted cells. The overlap penalty function approach was employed so as to prevent the disturbance of the placement of the remaining cells when performing an interchange of cells or a displacement of a single cell.

The normalization parameter p_2 is required since function $C_1()$ scales linearly with the grid size of the net list data, while function $C_2()$ is quadratic with respect to the grid size. If the normalization is not performed and the initial average values of the function $C_2()$ are substantially greater than the initial average values of the function $C_1()$, then the simulated annealing algorithm will primarily be concerned with the minimization of the overlap penalty function. That is, relatively little attention will be paid to the minimization of the TEIC. On the other hand, if the initial average values of $C_2()$ are substantially smaller than the average values of $C_1()$, then the simulated annealing algorithm will be concerned primarily with the minimization of the TEIC. That is, relatively little attention will be paid to the elimination of the cell overlaps, until T is relatively small. At this point, *most* of the attention will then be focused on the removal of the infeasibilities, with the accompanying increase in the TEIC.

Tests were performed on nine industrial circuits to determine the value of p_2 which would, on average, result in the lowest values of the final TEIC. The results showed that the lowest final values of the TEIC were obtained when

$$p_2\, C_2(\,) \;=\; \eta\, C_1(\,) \tag{4.6}$$

at the highest temperatures ($T = T_\infty$) and where $\eta \approx 0.5$. The performance of TimberWolfMC was not very sensitive to the value of η. No degradation in average performance was noted until η was reduced below 0.25 or beyond 1.0.

The implementation of $C_2()$ requires very many queries of the form: Given a cell $i \in \{1, \cdots, N_c\}$, where N_c is the number of macro/custom cells, find the set $\{J\}$ of other cells such that:

$$\{J\} \;=\; \{\, j \in \{1, \cdots, N_c\} \;\mid\; O_c(\,i\,,j\,) \neq 0 \,\} \tag{4.7}$$

Certainly by checking every other cell it can be determined which of these cells overlap with cell i. However, the *time complexity* (that is, the computation time as related to the number of cells), is $O(N_c)$. For circuits containing on the order of 30 cells or more, the time spent doing overlap computations can dominate the overall CPU time.

In the implementation of TimberWolfMC, the time complexity of overlap calculations was reduced by employing a *bin* storage technique. In this method, the core area is divided into a union of N_B nonintersecting rectangular bins of approximately the same size which fully cover the core area. Note that this implies that every (x,y) coordinate in the core area is contained by (or *hashes* to) exactly one bin.

The bins are numbered from 1 to N_B, such that:

$$N_B \;=\; N_{BX}\, N_{BY} \tag{4.8}$$

where the vertical columns of bins are numbered from 1 to N_{BX} and where the horizontal rows of bins are numbered from 1 to N_{BY}. Note that each vertical line given by $x = x_0$ is contained by exactly one of the columns of bins. Similarly, each horizontal line expressed by $y = y_0$ is contained by exactly one of the rows of bins. The N_B bins are stored in a two-dimensional array. The first index into the array corresponds to a particular vertical column of bins, and the second index corresponds to a particular horizontal row of bins.

A cell i is assigned to a specific bin in the following manner, where the function hash(a) returns the column, or row, index to which a particular coordinate a belongs:

```
find_bin( i ) {
        L' = L - ΔL ;
                /* L is the left edge of the bounding box of the cell, and ΔL is the amount of the out-
                ward expansion of the left edge due to the interconnect area estimated by the algorithm
                described in Chapter 6.  */
        c_l = hash( L' ) ;
                /* c_l is the column obtained by hashing.  */
        R' = R - ΔR ;
                /* R is the right edge of the bounding box of the cell, and ΔR is the amount of the out-
                ward expansion of the right edge due to interconnect-area estimation.  */
        c_r = hash( R' ) ;
        if( c_l ≠ c_r ) {
                return(0) ;
                        /* The cell is not contained by only one column of bins and hence it cannot pos-
                        sibly be contained by exactly one bin. That is, one or more bin boundaries are
                        overlapped by the cell. In this case, the cell is assigned to the bin 0, known as the
                        bucket.  */
        } else {
                        /* c_l = c_r, in which case the cell potentially could belong to a bin whose hori-
                        zontal (or first) index is given by c_l. The bottom and top edges must be checked
                        before it can be determined whether the cell is in fact contained by exactly one
                        bin.  */
                B' = B - ΔB ;
                        /* B is the bottom edge of the bounding box of the cell, and ΔB is the amount of
                        the outward expansion of the bottom edge due to interconnect area estimation.  */
                T' = T - ΔT ;
                r_b = hash( B' ) ;
                r_t = hash( T' ) ;
                if( r_b ≠ r_t ) {
                        return(0) ;
                } else {
                                /* r_b = r_t ; assign cell i to the bin represented by to the bin represented
                                by array_of_bins[ c_l , r_b ]  */
                        return( c_l + (r_b - 1) N_BX ) ;
                }
        }
}
```

When searching for the set $\{J\}$ of cells overlapping with cell i, it is necessary to search a number of bins given by:

$$\left(c_r - c_l + 1 \right)\left(r_t - r_b + 1 \right) + 1 \tag{4.9}$$

where the last 1 represents the search of the bucket. If the core is subdivided into very many bins, then the time to search a given bin is very short. However, the bucket will contain a large number of cells. On the other hand, if the core is subdivided into relatively few bins, the number of cells in each bin is relatively large, implying long search times since at least one large bin must be completely searched.

The goal is to find an expression yielding the optimal number of bins. The determination of the optimal number of bins is based on the following: N_B is optimal when the average number of cells in each bin is equal to the average number of cells in the bucket. A simple model was used to derive the optimal number of bins. In this model, it is assumed that the core area is square and that the cells to be placed into the core all have a square shape and that they all have the same size. It is assumed that the interconnect area associated with each cell edge has already been included in the determination of the cell boundaries. Let the number of cells be given by N_C, and assume that $\sqrt{N_c}$ is an integer. Thus the cells can be thought of as being stacked in columns of $\sqrt{N_c}$ cells as well as being arranged in rows of $\sqrt{N_c}$ cells.

It is further assumed that the number of columns of bins is equal to the number of rows of bins, that is,

$$N_{BX} = N_{BY} = \sqrt{N_B} \tag{4.10}$$

where it is also assumed that $\sqrt{N_B}$ is an integer. Note that the number of vertical bin partitions in the horizontal direction and the number of horizontal bin partitions in the vertical direction are each given by $(\sqrt{N_B} - 1)$. The bins are implemented such that each vertical (and horizontal) bin boundary (itself) belongs to exactly one bin.

Note that $\sqrt{N_c}$ cells cross each vertical bin boundary and $\sqrt{N_c}$ cells cross each horizontal bin boundary. These are the cells which must belong to the bucket. However, note that at each of the $(\sqrt{N_B} - 1)^2$ crossing points of the vertical and horizontal bin boundaries, a single cell is counted as having crossed both a vertical bin boundary as well as a horizontal bin boundary. Hence, a good approximation to the total number of cells which must be placed in the bucket (n_{B0}) may be expressed as:

$$n_{B0} = 2\sqrt{N_c}\left(\sqrt{N_B} - 1\right) - \left(\sqrt{N_B} - 1\right)^2 \tag{4.11}$$

The average number of cells per bin (c_b) is given by:

$$c_b = \frac{N_c}{N_B} \tag{4.12}$$

As stated above, the optimal number of bins is such that the expected number of cells in the bucket equals the expected number of cells per bin. That is, Eqn. 4.11 must equal Eqn. 4.12 .

$$\frac{N_c}{N_B} = 2\sqrt{N_c}\left(\sqrt{N_B} - 1\right) - \left(\sqrt{N_B} - 1\right)^2 \tag{4.13}$$

By defining $B = \sqrt{N_B}$, a simplification of Eqn. 4.13 yields:

$$B^4 - \left(2\sqrt{N_c} + 2\right)B^3 + \left(2\sqrt{N_c} + 1\right)B^2 + N_c = 0 \tag{4.14}$$

There are four possible roots for Eqn. 4.14, only one of which is applicable to the determination of the optimal number of bins. While it is possible to write down the four expressions for the roots of Eqn. 4.14, they are rather complicated. Furthermore, it cannot be determined which of the four expressions will give the realizable root for a given value of N_C. Since the expected value of N_C is less than 1000, and since the expected number of cells per bin is at least two, it is anticipated that the value of N_B would be less than approximately 500. That is, the expected value of B for the vast majority of macro/custom cell placement problems would not exceed 20 to 25. Since only integral values of N_B are of interest, the simplest way of solving Eqn. 4.14 is to start with $B = 0$ and increment B in steps of 0.1 until a zero crossing is noted for the left-hand side of Eqn. 4.14. Let the root of Eqn. 4.14, determined in this manner, be designated as B_O. The rounded value of the square of B_O then yields the desired number of bins, or N_B.

The values of N_{BX} and N_{BY} are then chosen as the best possible compromise subject to:

$$N_{BX} N_{BY} \approx N_B \tag{4.15}$$

and

$$N_{BX} \approx N_{BY} \tag{4.16}$$

The search time to determine the set of cells $\{J\}$ overlapping with a cell i is thus reduced by a factor of order:

$$\approx \frac{1}{N_B}$$

Of interest is the time complexity of this relationship, that is, the rate of growth of N_B with the number of cells (N_c). Since it is only of interest to find the rate of growth of N_B for *large* N_c, Eqn. 4.14 simplifies to:

$$B^4 - \left(2\sqrt{N_c}\right)B^3 + N_c = 0 \tag{4.17}$$

Following the strategy given in, Eqn. 4.17 has the *resolvent cubic equation* expressed below:[1]

$$y^3 + (-4\,N_c)\,y - 4\,N_c^2 = 0 \tag{4.18}$$

The solution (y) to Eqn. 4.18 is then used to compute R, defined by

$$R = \sqrt{N_c + y} \tag{4.19}$$

The following definitions are used:

$$D = \left\{ 3\,N_c - R^2 + 2\frac{N_c^{1.5}}{R} \right\}^{\frac{1}{2}} \tag{4.20}$$

and

[1] S. Selby, ed., "CRC Standard Mathematical Tables," (1973).

$$E = \left\{ 3N_c - R^2 - 2\frac{N_c^{1.5}}{R} \right\}^{\frac{1}{2}} \tag{4.21}$$

The four roots of Eqn. 4.17 are then given by:

$$B = -\frac{\sqrt{N_c}}{2} + \frac{R}{2} + \frac{D}{2} \tag{4.22}$$

$$B = -\frac{\sqrt{N_c}}{2} + \frac{R}{2} - \frac{D}{2} \tag{4.23}$$

$$B = -\frac{\sqrt{N_c}}{2} - \frac{R}{2} + \frac{E}{2} \tag{4.24}$$

$$B = -\frac{\sqrt{N_c}}{2} - \frac{R}{2} - \frac{E}{2} \tag{4.25}$$

The first step towards finding B as a function of N_c is to solve Eqn. 4.18 for y. The solution for Eqn. 4.18 is given by:

$$y = \left\{ 2N_c^2 + \sqrt{4N_c^4} \right\}^{\frac{1}{3}} + \left\{ 2N_c^2 - \sqrt{4N_c^4} \right\}^{\frac{1}{3}} \tag{4.26}$$

where only the real solution was of interest here, and where it was assumed that $N_c \gg 1$. Eqn. 4.26 can be simplified to:

$$y = \left\{ 2N_c \right\}^{\frac{2}{3}} \tag{4.27}$$

Furthermore, R can then be determined from Eqn. 4.19:

$$R = \left\{ N_c + (2N_c)^{\frac{2}{3}} \right\}^{\frac{1}{2}} \tag{4.28}$$

Rewriting Eqn. 4.28,

$$R = \sqrt{N_c} \left\{ 1 + \frac{(2N_c)^{\frac{2}{3}}}{N_c} \right\}^{\frac{1}{2}} \tag{4.29}$$

Making use of the approximation

$$\sqrt{1+x} \approx 1 + \frac{x}{2} \tag{4.30}$$

for $x \ll 1$, Eqn. 4.29 becomes

$$R = \sqrt{N_c} \left\{ 1 + \frac{1}{(2N_c)^{\frac{1}{3}}} \right\} \tag{4.31}$$

Substituting Eqn. 4.31 into Eqn. 4.20,

$$D = \left\{ 3N_c - N_c \left[1 + \frac{2}{(2N_c)^{\frac{1}{3}}} \right] + \frac{2N_c}{\left[1 + \frac{1}{(2N_c)^{\frac{1}{3}}} \right]} \right\}^{\frac{1}{2}} \tag{4.32}$$

where Eqn. 4.30 was again employed. Further simplification of Eqn. 4.32 yields

$$D = \left\{ 2N_c \left[1 - \frac{1}{(2N_c)^{\frac{1}{3}}} \right] + 2N_c \left[\frac{1}{1 + \frac{1}{(2N_c)^{\frac{1}{3}}}} \right] \right\}^{\frac{1}{2}} \tag{4.33}$$

Making use of the approximation:

$$1 + x \approx \frac{1}{1 - x} \tag{4.34}$$

for $x \ll 1$, and by using Eqn. 4.30, Eqn. 4.33 simplifies to

$$D = 2\sqrt{N_c} \left\{ 1 - \frac{1}{2} \left[\frac{1}{(2N_c)^{\frac{1}{3}}} \right] \right\} \tag{4.35}$$

A similar derivation starting with the substitution of Eqn. 4.31 into Eqn. 4.21 results in $E = 0$. Hence, Eqns. 4.24 and 4.25 yield negative solutions for B and must be discarded. Attention then focuses on Eqns. 4.22 and 4.23. Substitution of Eqns. 4.31 and 4.35 into Eqn. 4.22 yields:

$$B = \left\{ \frac{\sqrt{N_c}}{2} \left[\frac{1}{(2N_c)^{\frac{1}{3}}} \right] \right\} + \left\{ \sqrt{N_c} - \frac{\sqrt{N_c}}{2} \left[\frac{1}{(2N_c)^{\frac{1}{3}}} \right] \right\} \tag{4.36}$$

The simplification of Eqn. 4.36 yields $B = \sqrt{N_c}$. Substituting for B,

$$N_B \approx O(N_c) \tag{4.37}$$

The optimal number of bins is obviously not proportional to the number of cells. Attention then turns to the only choice remaining from Eqns. 4.22 through 4.25. Substitution of Eqns. 4.31 and 4.35 into Eqn. 4.23 yields

$$B = \left\{ \sqrt{N_c} + \frac{\sqrt{N_c}}{2} \left[\frac{1}{(2N_c)^{\frac{1}{3}}} \right] \right\} - \left\{ \sqrt{N_c} - \frac{\sqrt{N_c}}{2} \left[\frac{1}{(2N_c)^{\frac{1}{3}}} \right] \right\} \tag{4.38}$$

Simplifying,

$$B = \sqrt{N_c}\left[\frac{1}{(2N_c)^{\frac{1}{3}}}\right] \tag{4.39}$$

and thus

$$B \approx \frac{N_c^{\frac{1}{2}}}{N_c^{\frac{1}{3}}} \tag{4.40}$$

Finally, the relationship between the number of bins and the number of cells is the following:

$$N_B \approx O(N_c^{\frac{1}{3}}) \tag{4.41}$$

As a result of the bin technique, recall that the search time to determine the set of cells $\{J\}$ overlapping with a cell i is reduced by a factor of order

$$\approx \frac{1}{N_B}$$

Equation 4.41 implies that the comparative search time decreases as

$$O(N_c^{\frac{1}{3}})$$

The time complexity of the bin searches is also important. The search time is proportional to the average number of cells per bin, which varies with N_c as:

$$c_b \approx \frac{O(N_c)}{O(N_c^{\frac{1}{3}})} \tag{4.42}$$

or simply,

$$c_b \approx O(N_c^{\frac{2}{3}}) \tag{4.43}$$

Thus the time required to search each bin, and hence the time spent doing overlap calculations, grows as

$$O(N_c^{\frac{2}{3}}) \tag{4.44}$$

4.3.1.3 The third term in the cost function

The third part of the overall TimberWolfMC cost function, $C_3()$, is a penalty function which serves to limit the number of pins in each pin site to the capacity of that site. The custom cells are represented by j, where $j \in \{1, 2, \cdots, N_c^{'}\}$. The pin sites are represented by s_j, where $s_j \in \{1, 2, \cdots, N_s^j\}$, and where j is a specific custom cell. The function $C_t()$ returns the contents of a given pin site (that is, the number of pins contained by the site). Further, the function $C_p()$ returns the capacity of a given pin site.

The function $E_s(s_j)$ returns the penalty which is assigned to site s_j on custom cell j. A nonzero penalty is assigned to any site containing a number of pins which is greater than the specified site capacity. The expression for $E_s()$ is given below:

$$E_s(s_j) = \begin{cases} 0 & \text{if } C_t(s_j) \le C_p(s_j) \\ C_t(s_j) - C_p(s_j) + \kappa & \text{if } C_t(s_j) > C_p(s_j) \end{cases} \qquad (4.45)$$

The constant κ serves to ensure that the number of pin sites containing more pins than specified by the capacity goes to zero just prior to the end of stage 1 of TimberWolfMC. In the actual implementation, $\kappa = 5$.

With reference to Eqn. 4.45, the penalty function $C_3()$ is given by:

$$C_3() = \sum_{j=1}^{N_c'} \sum_{s=1}^{N_s^j} [E_x(s)]^2 \qquad (4.46)$$

4.3.2 The generate() function

4.3.2.1 Introduction

Details of the function responsible for the generation of new states for stage 1 are now presented. Tests have corroborated the theoretical results which have shown that the initial state has no influence on the final value of the TEIC.[1] Hence it is common to begin with a random initial placement of the macro/custom cells. New states can be generated in several possible ways. (1) A pair of cells (either could be a macro cell or a custom cell) are selected for interchange. (2) A pair of cells are selected for interchange in which both cells have their orientations changed prior to the assignment of their new locations. (3) A single cell is selected for displacement to a randomly chosen location. (4) A single cell is selected for a displacement to a new location in which the cell has its orientation changed prior to the assignment of its new location. (5) A single cell is selected to have its orientation changed in which the cell maintains its current location. (6) A custom cell is selected for an aspect ratio change. (7) An uncommitted pin group or sequence on a custom cell is selected and assigned to one or more new sites.

Each iteration of the *inner loop* initiates a call to the generation of new states function: generate(). The implementation of generate() is now presented.

---

[1] D. Mitra, F. Romeo, and A. Sangiovanni-Vincentelli, "Convergence and Finite-Time Behavior of Simulated Annealing," (1985).

```
generate ( ) {
    p = R_i( 1 , 2 , p ) ;
        /* R_i( 1 , 2 , p ) returns a random integer between 1 and 2, such that the probability of
        returning 1 is p and the probability of returning 2 is (1 – p). The value r = p / (1 – p)
        represents the ratio of single-cell displacements to pairwise interchanges. */
    if ( p == 1 ) {
        i = R( 1 , N_c ) ;
        x = R( c_l , c_r ) ;
        y = R( c_b , c_t ) ;1
            /* R(k,l) returns a random integer between k and l, with uniform distribution. A
            cell i is randomly selected for displacement to a randomly selected new location,
            represented by the coordinate pair (x,y) in the core area (bounded on the left by c_l,
            on the right by c_r, below by c_b, and above by c_t). The coordinate pair (x,y) rep-
            resents the new location for the center of cell i. */
        if( A_1( i , x , y ) == YES ) {
            /* Displacement of cell i to (x,y) was accepted. */
        } else if( A'_1( i , x , y ) == YES ) {
            /* Cell i is the focus of another new state attempt; this time its orientation is
            changed to effect an aspect ratio inversion. In Figure 4.5, if cell C_2 is displaced
            to location (x,y) with its current orientation, overlap will result with the core
            boundary and with cell C_3. If the orientation of cell C_2 is changed such that its
            aspect ratio is inverted, no overlap will result in the new location. */
        } else {
            A'_0( i ) ;
                /* The displacement to (x,y) failed for both aspect ratios; the next at-
                tempt is a randomly chosen orientation change of cell i. */
        }
        if( cell i is a custom cell ) {
            for( k = 1 ; k ≤ P_U(i) ; k++ ) {
                A_p(i) ;
                    /* P_U(i) returns the number of uncommitted pins on cell i. This
                    value dictates the number of times a new pin group (or sequence)
                    is selected for displacement to a new set of pin sites. */
            }
            A_r(i) ;
                /* Attempt an aspect ratio change, within specified bounds. */
        }
```

```
        } else {          /* Two cells will be selected for interchange. */
                j = R( 1 , N_c ) ;
                if( A_2( i , j ) == YES ) {
                        /* A_2 returns YES if the interchange of cells i and j was accepted. */
                } else {
                        A'_2( i , j ) ;
                                /* An interchange of cells i and j will be attempted again; this time the
                                orientations of the cells are each changed to effect aspect ratio inver-
                                sions. For example, in Figure 4.5, if cells C_5 and C_7 are interchanged,
                                a great deal of overlap will result with the core boundary and with cells
                                C_4 and C_6. However, if the orientation of each cell is changed such
                                that its respective aspect ratio is inverted, no overlap will result in their
                                new locations. */
                }
        }
}
```

The ratio r of single cell displacements to cell interchanges can have an important effect on the final TEIL. Table 4.1 shows the results for one particular circuit obtained from Gould-AMI Semiconductors containing 25 macro cells.

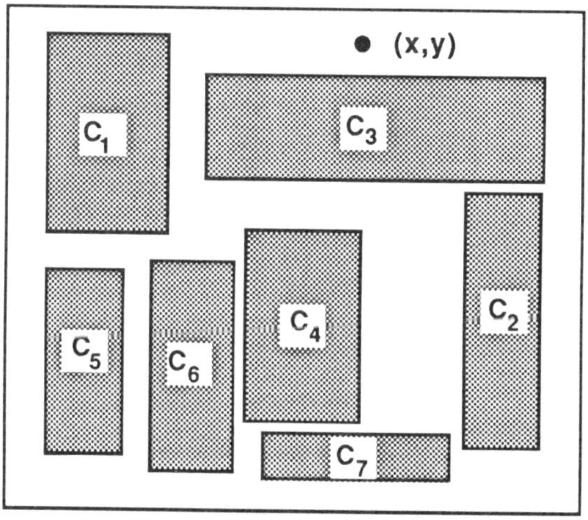

Figure 4.5 The displacement of cell C_2 to location (x,y) is much more likely to be accepted if its orientation is changed such that its aspect ratio is inverted. Also, if cells C_5 and C_7 are to be interchanged, inverting their respective aspect ratios greatly increases the acceptance probability.

Ratio *r*	Average Final TEIL	No. of Trials
1	171,378	4
3	165,152	4
7	162,033	4
15	160,456	4
31	171,898	4

Table 4.1 The effect of the ratio *r*.

The cell placement generated by AMI's CIPAR placement and routing package yielded a TEIL value of 237,835. The TimberWolfMC placement data was obtained with $\alpha = 0.90$ and with $A_c = 100$. Table 4.1 also indicates that a value of r in the range of 7 to 15 yields the best overall results. The implementation in TimberWolfMC has r set to 10.

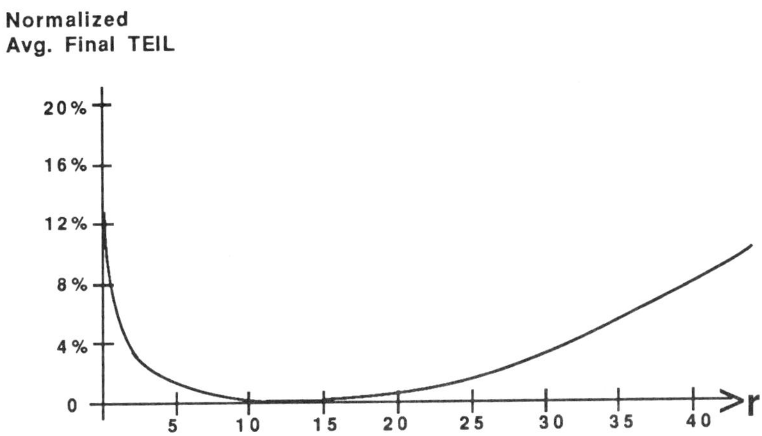

Figure 4.6 The normalized final TEIL for a variety of circuits versus the ratio r of single-cell displacements to cell interchanges.

The normalized final TEIL for a variety of circuits versus the ratio r is shown in Figure 4.6. The circuits contained an average of about 25 macro cells. Each execution of the inner loop consisted of $A_c = 200$ calls to the generate() function per cell. The initial value of $T = T_\infty$ was such that almost every new state was accepted. For each successive iteration of the inner loop, the new value of T was determined by: $T_{new} = \alpha T_{old}$, where $\alpha = 0.90$. Figure 4.6 indicates that values of r in the range from 7 through 15 yield TEIL values which are within one percent of the minimum. This wide range gives robustness to the algorithm.

4.3.2.2 The Range Limiter

The generate() function is controlled by a *range limiter*. Large-distance moves usually imply large values of ΔC. At low temperatures, only moves which approximately satisfy $\Delta C \leq 0$ have a reasonable chance of being accepted. Hence, at low temperatures, the large-distances moves are almost invariably rejected. In order to generate moves which have a reasonable probability of acceptance, these large-distance moves are therefore prohibited by the use of a range-limiter window. When a cell is selected for displacement, such as cell C_1 (whose bounding box is centered at (x_0, y_0) in Figure 4.7), the randomly selected new location for the center of the bounding box of the cell must lie within the range-limiter window centered at (x_0, y_0). In the current implementation of TimberWolfMC, cell interchanges are not controlled by the range-limiter window.

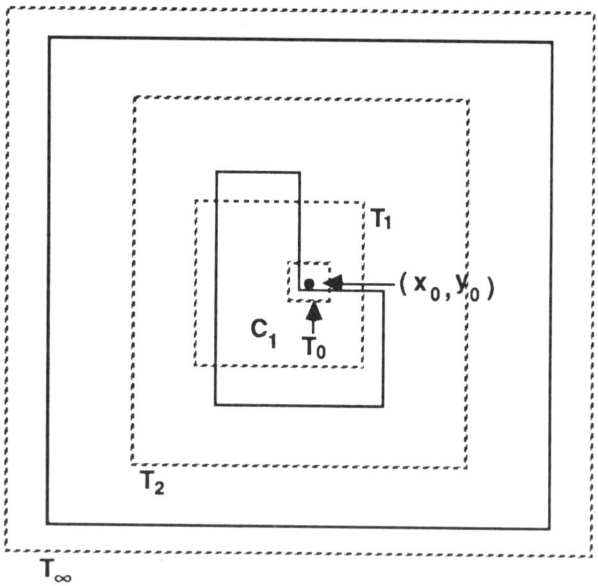

Figure 4.7 The span of the range-limiter window is proportional to log(T). At $T = T_\infty$, the window has maximum span, and at $T = T_0$, the window has minimum span.

At the beginning of stage 1, when T is at its maximum value (indicated as T_∞ in Figure 4.7), the window extends beyond the core area, thereby allowing moves of maximum distance. Values of T are considered over a range of approximately six decades. As T is lowered (for example, to T_2 and subsequently to T_1), the window span in each direction is reduced as a function of the base-ten logarithm of T. In Figure 4.7, the span of the window is indicated for two intermediate values of T, namely T_1 and T_2. As T approaches the value $T = T_0$, the span of the window reaches its minimum value (as indicated in Figure 4.7) and this condition marks the end of stage 1 of TimberWolfMC.

Two different range-limiting functions were tested. For the first case, the span of the window was proportional to the logarithm of T. That is, the window span linearly decreases with the logarithm of T. The actual expressions controlling the respective window dimensions are shown below:

$$W_x(T) = W_x^{\infty}\left\{ \varepsilon \frac{\log(T)}{\log(T_{\infty})} \right\} \tag{4.47}$$

$$W_y(T) = W_y^{\infty}\left\{ \varepsilon \frac{\log(T)}{\log(T_{\infty})} \right\} \tag{4.48}$$

where W_x^{∞} represents the horizontal span of the chip and W_y^{∞} represents the vertical span of the chip.

ε	Average Final TEIL Reduction
20	11%
5	13%
0.1	17%
0.01	16%

Table 4.2 The effect of ε on the range limiter.

Experiments were performed to determine the best value for ε. The results for one of the circuits tested, a 6-cell Gould-AMI circuit, are shown in Table 4.2. The number of executions of generate() per cell per value of T (or A_c) was 20. The initial value of T was $T_{\infty} = 10^5$. The results shown in Table 4.2 indicate that $\varepsilon \approx 0.1$ yields the best results. Larger values of ε apparently result in the generation of more large-distance moves which have correspondingly smaller chances of being accepted. Also, not enough shorter range moves are generated.

A second (nonlinear) range-limiter function was also tested. This function is characterized by a more rapid decrease of span for the higher values of T and a slower rate of decrease for the lower values of T.

$$W_x(T) = W_x^{\infty}\left\{ \frac{\rho^{\log(T)}}{\lambda} \right\} \tag{4.49}$$

and

$$W_y(T) = W_y^{\infty}\left\{ \frac{\rho^{\log(T)}}{\lambda} \right\} \tag{4.50}$$

The value of λ was chosen such that for $T = T_{\infty}$, the term in the braces on the right hand sides of Eqns. 4.49 and 4.50 are normalized to 1.0. That is,

$$\lambda = \rho^{\log(T_{\infty})} \tag{4.51}$$

A nominal value of T_{∞} is 10^5.

The smallest values of the final TEIL were achieved with ρ in the range $1 \leq \rho \leq 4$. For this range, there was no observable difference in the performance with respect to the final TEIL. However, smaller values of the *residual cell overlapping* after the completion of the stage 1 algorithm were achieved with the larger values of ρ. The residual cell overlapping is the value of the penalty function $C_2()$ when $T \rightarrow T_0 \approx 0$. For a given value of T, as ρ increases, the window size is smaller. This implies that more local cell moves will be attempted, thus enhancing the probability of the removal of instances of cell overlapping. Consequently, the choice of $\rho = 4$ was made to obtain the lowest possible values of both the final TEIL and the residual cell overlapping for the second range limiter.

Table 4.3 summarizes the results of a performance comparison of the two range limiters. Similar behavior was noted for several other industrial circuits.

Range Limiter	Average Final TEIL	Average Final Residual Overlap	Trials
Eqns. 4.49/4.50	162,668	93	4
Eqns. 4.47/4.48	162,851	2717	4

Table 4.3 A comparison of the two range-limiting functions.

The circuit used for the comparison in Table 4.3 contained 25 macro cells and was from Gould-AMI. The cell placement generated by AMI's CIPAR placement and routing package yielded a final TEIL of 237,835. The simulated annealing parameters included: $T_\infty = 10^5$, $\alpha = 0.90$ and $A_c = 100$. Approximately 125 values of T were used, for a total of approximately 300,000 executions of generate() per trial. The conclusions drawn from the data in Table 4.3 are that while there is no difference in the performance of the two range limiters in terms of final TEIL, the range limiter expressed by Eqns. 4.49 and 4.50 yields a noticeably smaller value of residual cell overlapping after the completion of the stage 1 algorithm. Hence, the form for the range limiter expressed by Eqns. 4.49 and 4.50 was selected for implementation in TimberWolfMC.

4.3.2.3 Single-cell displacement-point selection

In the function generate(), when a particular cell is selected for displacement, the new target location is a randomly selected coordinate pair within the range-limiter window. The process of selecting the new target location is performed by the *displacement-point selection* function. Two different functions were tested: (1) The first displacement-point selection function is $D_s()$, (also referred to as the *discrete stepper*) which selects only from a very limited number of evenly dispersed points lying within the range-limiter window. (2) The second displacement-point selection function is $D_r()$, which selects randomly from any of the points lying within the range-limiter window.

The function $D_s()$

In this method, a cell selected for displacement can be moved to only a very limited number of points. The value of $0.5 \cdot W_x(T)$ defines the maximum distance in x that a cell's center may be moved in either direction. The function $D_s()$ restricts the step sizes in each direction to a set of 3 equally separated amounts. The step sizes in each direction are multiples of s_x, defined as follows:

$$s_x = \frac{W_x(T)}{6} \tag{4.52}$$

and similarly for the y direction:

$$s_y = \frac{W_y(T)}{6} \tag{4.53}$$

The discrete stepper only permits a step in the x direction which is an integral multiple of s_x. Furthermore, a step in the y direction must be an integral multiple of s_y. Therefore, the only possible integers that can be multiplied by s_x (or s_y) to yield the actual step distance are contained in the set $I_x = I_y = \{-3, -2, -1, 0, 1, 2, 3\}$. Other integers, when multiplied by s_x, would yield step distances which extend beyond the boundaries of the range-limiter window. Excluding the selection $I_x = 0$ and $I_y = 0$, there are 48 possible points within the range-limiter window which could be selected as the displacement points (that is, the points to which the center of the selected cell could be placed). As T approaches the value T_0, the span of the window reaches its minimum value of 6 units (in terms of the gridsize inherent in the specification of the cell geometry and pin locations in the netlist) and this condition marks the end of stage 1 of TimberWolfMC. Thus, the minimum value of the step sizes s_x and s_y are one.

The function $D_s()$ has the following two characteristics: (1) With far fewer points to be considered, it is likely that a displacement of a cell to most of the 48 points could be considered for each value of T. (2) Much more emphasis is placed on the cells making large moves during large values of T and on the cells making comparatively smaller moves for smaller values of T. That is, small refinement moves are not attempted at high values of T, nor are large moves attempted at low values of T.

The comparative performance of $D_s()$ versus $D_r()$

For the function $D_r()$, a cell selected for displacement can be moved to any of the points within the range-limiter window. TimberWolfMC placement results comparing the two methods of displacement-point selection are presented in Tables 4.4 and 4.5.

The circuit used for the comparisons was from Gould-AMI. This circuit contained 25 macro cells and 17 pads, as well as 161 nets and a total of 413 pins. The cell placement generated by AMI's CIPAR placement and routing package yielded a total estimated interconnect length of 237,835.

The values of some of the simulated annealing parameters included: $T_\infty = 10^5$ and $\alpha = 0.90$. The results were tabulated for a wide range of values for A_c. The effect of A_c on the final results is also readily apparent from these tables. The computation time was, for example, 4.0 hours on a VAX 8600 for $A_c = 512$. The computation time is approximately linearly related to the value of A_c.

As described in Section 4.2.2, the size of the core area is a function of the expected interconnect area for the circuit layout. Furthermore, since the computation of the interconnect area is a function of the average interconnection length for a series of random cell placements, there is a small variance in the core area from trial to trial. It would be expected that the amount of residual cell overlapping after the completion of stage 1 would be a function of the magnitude of the core area. That is, if the core area is larger, the algorithm has more free areas available in trying to remove overlaps. On the other hand, a larger core area implies that the cells are more likely to be placed somewhat farther apart than for a smaller core area. This would lead to the expectation of a greater final TEIL for larger core areas. The experiments corroborated these expectations.

A_c	Avg. Final TEIL $D_s()$	Avg. Final TEIL $D_r()$	Avg. Final Overlap $D_s()$	Avg. Final Overlap $D_r()$	Trials $D_s()$	Trials $D_r()$
4	192,903	198,791	8756	10000	10	3
8	190,271	183,370	3134	5195	7	7
16	179,820	180,783	3740	5245	8	3
32	169,580	169,210	1092	1877	7	7
64	165,487	175,854	1003	1581	5	3
128	164,419	168,339	996	390	7	6
256	161,604	154,536	655	144	9	4
512	158,107	156,740	78	868	7	3
Avg.	172,773	173,452	2431	3163		

Table 4.4 Comparison of $D_s()$ versus $D_r()$ for a Core Area of 3129 × 3129.

Ac	Avg. Final TEIL $D_s()$	Avg. Final TEIL $D_r()$	Avg. Final Overlap $D_s()$	Avg. Final Overlap $D_r()$	Trials $D_s()$	Trials $D_r()$
4	196,280	191,640	7654	8465	16	8
8	188,113	182,233	2963	4967	16	8
16	174,703	180,042	3408	4775	20	8
32	166,503	168,997	1124	1964	12	8
64	164,529	170,969	1256	1007	12	8
128	164,419	164,954	996	948	7	8
256	161,292	153,800	678	224	16	8
512	154,799	154,257	127	805	12	8
Avg.	171,329	170,861	2276	2894		

Table 4.5 Comparison of $D_s()$ versus $D_r()$ for a variable core area.

Table 4.4 presents the TimberWolfMC placement results for the case in which each trial had the same core area of (3129 × 3129). Note that the function $D_s()$ performed at least as well as $D_r()$ for $A_c \in \{4, 16, 32, 64, 128\}$. The average final TEIL for all of the trials (independent of A_c) using the second method was 173,452. Meanwhile, the average final TEIL for all of the trials using $D_s()$ was 172,773, which represents an insignificant improvement over the second method. The average residual cell overlapping for all of the trials using $D_r()$ was 3163. On the other hand, the average residual cell overlapping for $D_s()$ was 2431, which is 23 percent less than the average for $D_r()$. Note that for $D_r()$, in contrast to $D_s()$, each doubling of A_c failed, in some cases, to produce a further reduction in the average final interconnect length as well as a further reduction in the average amount of residual cell overlapping.

Note further the effect of A_c on the average final TEIL for the function $D_s()$. The average final TEIL for $A_c = 512$ is 18 percent smaller than the result for $A_c = 4$. Of course, the run time for $A_c = 512$ is nearly 100 times longer than for $A_c = 4$. It is also apparent that each doubling of A_c produces a further reduction in the average final TEIL as well as a further reduction in the average amount of residual cell overlapping.

The results in Table 4.5 were accumulated for three different core sizes: 2945 × 2945, 3129 × 3129, and 3313 × 3313. Note that the function $D_s()$ performed at least as well for $A_c \in \{16, 32, 64, 128\}$. The average final TEIL for all of the trials using $D_r()$ was 170,861. Meanwhile, the average final TEIL for all of the trials using $D_s()$ was 171,329 . The average residual cell overlapping for all of the trials using $D_r()$ was 2894. On the other hand, the average residual cell overlapping for all of the trials using $D_s()$ was 2276, which is 21 percent less than the average for $D_r()$.

TimberWolfMC uses $D_s()$ for the following reasons: (1) While $D_s()$ and $D_r()$ appear to yield approximately the same final TEIL overall, $D_s()$ appears to perform slightly better for the middle range values of A_c (that is, 16, 32, 64, and 128). For values of $A_c \approx 48$ (the number of possible points within the range-limiter window which could be selected as the displacement points), the function $D_s()$ has a slight edge in performance over $D_r()$. This is likely due to the ability of $D_s()$ to sample more uniformly all of the regions within the range-limiter window. For values of $A_c \gg 48$, one would not expect to observe a significant difference in performance for the two methods (in fact, arguments could be made for $D_r()$). (2) $D_s()$ yields slightly lower values for the residual cell overlapping. (3) $D_s()$ appears to yield a slightly more consistent behavior in the respect that each doubling of A_c produces a further reduction in the average final TEIL and usually a further reduction in the average amount of residual cell overlapping.

4.3.3 Additional stage 1 simulated annealing algorithmic details

The *inner loop criterion* in TimberWolfMC, that is, the number of iterations of the inner loop, is specified in terms of the number of attempted new states per cell per value of T (or A_c). The number of iterations of the inner loop for a given value of T is then given by:

$$A = A_c N_c \tag{4.54}$$

where N_c is the number of cells.

The effect of the inner loop criterion was noted for several circuits containing 30 to 60 macro cells, In Figure 4.8, a plot of the normalized average final TEIL versus A_c for these circuits is shown. Note that for circuits of this size, $A_c \approx 400$ is sufficient to yield the best results.

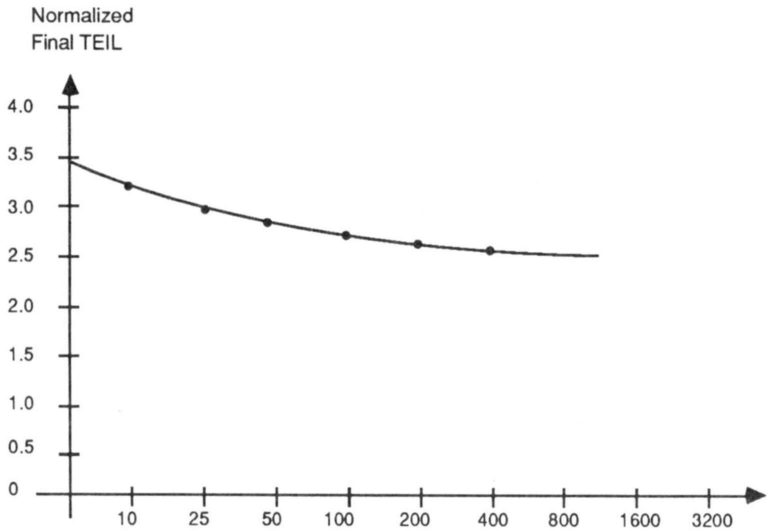

Figure 4.8 Normalized final TEIL versus A_c.

In Figure 4.9, a plot of the relative final chip area (following TimberWolfMC global routing and placement refinement) versus A_c for these circuits is shown. Again, $A_c \approx 400$ is sufficient to yield the best results. Also note that reasonably good results can be obtained for substantially smaller values of A_c. The execution time for stage 1 of TimberWolfMC is directly proportional to A_c. The CPU time for a typical $A_c = 400$ run was about 4 hours on an VAX 8650. Note that for $A_c = 25$, the reduction in TEIL was 13 percent less than the best results. However, the CPU time in this latter case was 16 times less than that for the $A_c = 400$ tests. While the $A_c = 400$ case represents a substantial amount of CPU time, the additional 13 percent reduction in TEIL often yields an additional final chip area reduction of as much as 10 to 15 percent. From this standpoint, in the latter stages of the circuit design process, the additional CPU time is usually justified. On the other hand, in the early stages of the design, a smaller value of A_c is more appropriate.

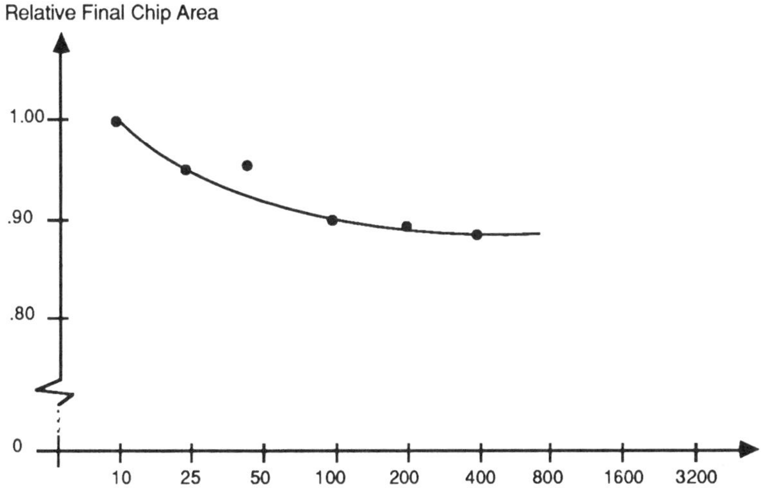

Figure 4.9 Relative final chip area versus A_c.

The function update(T) is expressed by

$$T_n = \text{update}(T) = T \cdot \alpha(T), \qquad 0 < \alpha(T) < 1 \tag{4.55}$$

where T_n is the new value of the temperature derived from the current value T. The cooling schedule used in TimberWolfMC was determined experimentally, with the following observations: (1) Three to five iterations are performed in which virtually every new state was accepted and where T is reduced rapidly from iteration to iteration. (2) After having left the high T regime, T was reduced such that the reduction in the average TEIC from one T step to another is approximately the same. (3) When the average value of the TEIC is only slightly reduced from one T step to another, the control of T enters the final regime where T is reduced more rapidly so that the value of the cost function converges.

The initial value of $T = T_\infty$ is chosen such that virtually every new state proposed by generate() is accepted. Tests on 9 industrial circuits showed that the average cell area $\bar{c}_a$, including the expected interconnect area, was approximately proportional to the value of $T = T_\infty$ necessary to obtain an initial acceptance rate of nearly 100 percent. For a 25-cell circuit characterized by $\bar{c}_a = 10^4$, the smallest value of T yielding an initial acceptance rate of nearly 100 percent was T_∞ approx 10^5. Therefore, to normalize T_∞ for different circuit sizes and different grid sizes, the following expression is used:

$$T_\infty = T_\infty^* \frac{\bar{c}_a}{\bar{c}_a^*} \tag{4.56}$$

where $\bar{c}_a^* = 10^4$ and $T_\infty^* = 10^5$ were the characteristics of the 25-cell industrial circuit. Defining

$$S_T = \frac{\bar{c}_a}{\bar{c}_a^*} \tag{4.57}$$

it is then possible to rewrite Eqn. 4.56 as:

$$T_\infty = S_T \cdot T_\infty^* \tag{4.58}$$

The value of S_T is a scale factor for the simulated annealing temperature profile, taking into account the circuit and grid sizes.

The temperature is updated by the function update() as follows:

$$T_n = \text{update}(T) = T \cdot \alpha(T), \qquad 0 < \alpha(T) < 1 \tag{4.59}$$

Table 4.6 contains the data for $\alpha(T)$ as a function of T for which the best results were obtained with TimberWolfMC.

For $T \geq$	$\alpha(T)$
$7000S_T$	0.85
$200S_T$	0.92
$10S_T$	0.85
0	0.80

Table 4.6 The TimberWolfMC cooling schedule, $\alpha(T)$ versus T.

The entries in the first column indicate the smallest value of $S_T{\cdot}T$ for which $\alpha(T)$ returns the corresponding entry in the second column. That is, for $T \geq 7000S_T$, $\alpha(T)$ returns 0.85, for $200S_T \leq T < 7000S_T$, $\alpha(T)$ returns 0.92, and so on. It was arbitrarily determined that approximately 120 temperature values were to be considered in a typical execution of the simulated annealing algorithm. Based on extensive experimentation for 9 industrial circuits, the range of T from $200S_T$ to $7000S_T$ influenced most strongly the performance of TimberWolfMC. Consequently, as many temperature points as possible were placed in this critical range while arbitrarily electing not to reduce T more rapidly than by a factor of 0.85. The performance of TimberWolfMC did not appear to be very sensitive to the temperature profile outside the critical range. The *stopping criterion* is satisfied following an iteration of the inner loop in which the range-limiter window has achieved its minimum span.

4.4 The Algorithms for Stage 2 of TimberWolfMC

The second stage of TimberWolfMC consists of several executions of a placement-refinement algorithm. Each execution consists of three steps: (1) a channel definition step, (2) a global routing step, and (3) a placement-refinement step. The information obtained in step 2 is used to compute the density of all of the channels, which then permits accurate interconnect-area determination. The placement of the cells is then refined in step 3 to reflect the required interconnect area. A low-temperature simulated annealing algorithm is used to accomplish step 3. Three such placement-refinement steps have proven to be sufficient for the final TEIL and the final chip area to converge. The placement of the cells requires very little modification during detailed routing as a result of this strategy. This section presents the three steps of the placement-refinement algorithm.

4.4.1 Channel generation

A new channel generation scheme was devised which emphasizes the creation of channels which are more suitable for placement refinement in comparison to those generated by traditional channel definition algorithms. Chapter 7 is devoted to a complete description of the new channel generation algorithm. A channel is created between every pair of cell edges such that: (a) the two edges belong to different cells, (b) the two edges are parallel, (c) the two edges overlap in one (of the two) dimensions (that is, a rectangular region of *empty space* is fully bounded on two sides by these two edges), and (d) no other cell edges enter the bounded region between the two edges. In addition, the channel generator permits cells to be of any rectilinear shape. This enables TimberWolfMC to monitor more carefully local routing congestion. In particular, it enables accurate determination of the space necessary between nearby cells. Traditional routing-channel generators have been designed only to facilitate detailed routing of the chip, and are less able to monitor local routing congestion.

4.4.2 Global routing

A new global routing methodology, described fully in Chapter 8, was also developed. This strategy is based on Lawler's algorithm for finding the M-shortest paths between two vertices on a graph.[1] Lawler's algorithm was used in the development of an algorithm for finding the approximately M-shortest paths on a graph for an n-pin ($n \geq 2$) net, with the ability to handle electrically equivalent pins. The new global router is also applicable to situations in which the cell placement is fixed, with fixed channel capacities. For this case, a random interchange algorithm was devised which selects, from the M alternatives for each net, the overall global routing which satisfies the capacity constraints and minimizes the total interconnect length.

For the case in which channel capacities are not fixed, the shortest routing is selected for each net. If M is made sufficiently large, it has been found in practice that the global router usually finds the minimal Steiner route for each net.

The global routing information is then used to compute the density of all of the channels. An efficient channel router has been developed which is able to route channels in a number of tracks which is almost invariably bounded by the density plus one.[2] The product of the user-supplied center-to-center track spacing and the density plus one determines the refined estimate of the interconnect space needed for each channel.

---

[1] E. Lawler, "Combinatorial Optimization: Networks and Matroids," (1976).

[2] J. Reed, A. Sangiovanni-Vincentelli, and A. Santamauro, "A New Symbolic Channel Router: YACR2," (1985).

4.4.3 Placement refinement

A low-temperature simulated annealing algorithm is used to accomplish this step. The placement of the cells is refined to reflect the required interconnect area, as determined by the channel definition and global routing steps. These steps determined the required interconnect area of every channel. Since exactly two cell edges border each TimberWolfMC channel, the spacing requirement between the two cell edges is immediately established. One half of the interconnect space is associated with each of the two cell edges. That is, each respective edge is expanded outward (from the interior of the cell) by an amount equal to one-half of the required channel width. These expanded cell edges are used each time the amount of overlap (if any) is updated for two cells. Note that in contrast to the stage 1 dynamic interconnect-area estimator, the amount of outward expansion of the cell edges is a static quantity for each execution of the three placement-refinement steps.

The same objective function as in stage 1 of TimberWolfMC, described in Section 4.3.1, is used. The function generate() has a much simpler form for stage 2. New states can only be generated by attempting single cell displacements and by alterations of the pin placement. The orientations of the cells are not changed during stage 2. The aspect ratios of custom cells also remain fixed during the second stage. These latter two methods of new state generation tend to invalidate the calculation of the interconnect area associated with a particular cell edge. More placement-refinement steps would then be required to achieve a convergence of the final chip area and the final TEIL. Furthermore, even if convergence is achieved, the final values of the chip area and TEIL tend not to be better than the case in which orientation and aspect-ratio changes are not permitted.

Each iteration of the inner loop initiates a call to the generation of new states function: generate(). Following the notation of Section 4.3.2, the implementation of generate() is now presented.

```
generate() {
        i = R(1, N_c) ;
        x = R( c_l , c_r ) ;
        y = R( c_b , c_t ) ;
        A_1( i, x, y ) ;
        if( i ∈ {1, ··· , N'_c } ) {
                        /* Cell i is a custom cell. */
                for( k = 1 ; k ≤ P_U(i) ; k++ ) {
                        A_p(i) ;
                                /* The function P_U(i) returns the number of uncommitted pins on cell
                                i. The returned value dictates the number of times a new pin group (or
                                sequence) is selected for displacement to a new set of pin sites. */
                }
        }
}
```

Since the cell moves are very local in stage 2, the range-limiter window must have an initial size which is some fraction μ of the span of the core area. Equation 4.49 dictates the control of the window size for the x direction. It is desired that:

$$\mu = \frac{W_x(T)}{W_x^{\infty}} \tag{4.60}$$

The initial value of T for stage 2 is selected which satisfies Eqn. 4.60. That is, a solution for T' is desired from the following expression:

$$\mu = \frac{4^{\log (T')}}{\lambda} \tag{4.61}$$

where λ is defined by Eqn. 4.51. That is,

$$\mu = \frac{4^{\log (T')}}{4^{\log (T_\infty)}} \tag{4.62}$$

Solving this expression for T' then yields:

$$T' = \mu^{\log_4 10}\, T_\infty \tag{4.63}$$

In TimberWolfMC, $\mu = 0.03$ is used. This implies that the cells are initially permitted to move 3 percent of the span of the core in any direction. Based on tests for 9 industrial circuits, on the average it was found that larger values of μ yielded equally good final values of the TEIL, but not better. Of course, the CPU time required to achieve a convergence in the TEIL is larger for larger values of μ. The value of μ had to be reduced below 0.03 before the average performance showed a degradation.

The temperature is updated by the function update() as follows:

$$T_n = \text{update}(T) = T \cdot \alpha(T), \qquad 0 < \alpha(T) < 1 \tag{4.64}$$

Table 4.7 contains the data for $\alpha(T)$ as a function of T which is used during stage 2.

For $T \geq$	$\alpha(T)$
$10S_T$	0.82
0	0.70

Table 4.7 The stage 2 cooling schedule, $\alpha(T)$ versus T.

The parameter S_T, defined by Eqn. 4.57, normalizes the temperature profile for various circuit and grid sizes. The entries in the first column indicate the smallest value of $S_T \cdot T$ for which $\alpha(T)$ returns the corresponding entry in the second column. That is, for $T \geq 10S_T$, $\alpha(T)$ returns 0.82 and for $0 < T < 10S_T$, $\alpha(T)$ returns 0.70.

For the first two iterations of the three placement-refinement steps, the stopping criterion is satisfied once an iteration of the inner loop has been performed with the range-limiter window at its minimum span. For the third and final iteration of the placement-refinement step, the stopping criterion is satisfied if the value of the cost function is unchanged for 3 consecutive iterations of the inner loop.

4.5 TimberWolfMC Results

The performance of TimberWolfMC was compared versus various industrial, university, and manual placement methods. The two criteria were the TEIL and the final chip area. The results for 9 industrial circuits of various sizes are summarized in Table 4.8.

Circuit	No. Cells	No. Nets	No. Pins	Total Interconnect	Area ($x \times y$)	Interconnect Red. (%)	Area Red. (%)
i1	33	121	452	7431	236×223	26	14
p1	11	83	309	12306	293×294	8	18
x1	10	267	762	60326	875×744	11	15
i2	23	127	577	121386	2873×2751	49	†
i3	18	38	102	7043	644×699	46	56
l1	62	570	4309	254063	1084×1042	19	50
d2	20	656	1776	419608	1355×1433	13	4
d1	17	288	837	37365	245×305	23	†
d3	17	136	665	325457	3398×3298	29	31
Avg.						24.9	26.9

†comparative result not available

Table 4.8 TimberWolfMC performance comparisons versus industrial, university, and manual placements.

The 26 percent reduction in TEIL and the 14 percent chip area reduction achieved for circuit i1 was in comparison to a placement method based on resistive network optimization.[1] Circuits i2 and i3 were in comparison to the CIPAR placement and routing package developed by Gould-AMI. The CIPAR router was not able to route its own placement for circuit i2, hence an area comparison was not possible.

The 19 percent total interconnect length reduction and the 50 percent chip area reduction was in comparison to a manual layout performed at Intel Corp. Circuit p1 was in comparison to a manual layout at Hewlett-Packard. Further, circuits d1, d2, and d3 were in comparison to manual layouts at Advanced Micro Devices. For circuits l1, p1, d1, d2, and d3, the widths of the power and ground lines were not supplied. It was assumed that, on average, the widths of these lines were about twice a normal wire width, and that these lines were present in every channel. The CPU time ranged from 15 minutes for the smallest circuits to 4 hours for the largest circuits on a DEC MicroVAX II.

4.6 Conclusion

A macro/custom cell placement and chip-planning package based on the simulated annealing algorithm has been developed. A new algorithm for accurately estimating the in-

[1] C. K. Cheng and E. Kuh, "Module Placement Based on Resistive Network Optimization," (1984).

terconnect area around the individual cells is used in TimberWolfMC. This dynamic interconnect-area estimation algorithm has resulted in the generation of placements which require virtually no placement modification during or after detailed routing.

The placement algorithm proceeds in two distinct stages. During the first stage, the interconnect area around the individual cells is determined using the dynamic interconnect-area estimator. That is, as each cell is moved, its effective area is adjusted. A simulated annealing algorithm is used to minimize the TEIC. The second stage of TimberWolfMC consists of several executions of a placement-refinement algorithm. Each execution consists of three steps: (1) a channel definition step, (2) a global routing step, and (3) a placement-refinement step. The information obtained in step 2 is used to compute the density of all of the channels, which then permits accurate interconnect-area determination. The placement of the cells is then refined in step 3 to reflect the required interconnect area.

The performance of the package was compared for 9 industrial circuits versus a variety of other placement algorithms. The average total interconnect length reduction was over 24 percent and the average chip area reduction was nearly 27 percent. Furthermore, for the 9 examples, circuit-area reductions ranged from 4 to 56 percent versus a variety of other placement algorithms.

Chapter 5

Average Interconnection Length Estimation

5.1 Introduction

This chapter presents models and algorithms for accurately estimating the total interconnect length for both random and optimized placements. The algorithms developed in the next chapter for accurately estimating interconnect *area* require a good estimate of the *final* total interconnect length. The final total interconnect length is determined by the macro/custom cell placement algorithm which was presented in the previous chapter. This placement algorithm was based on the simulated annealing algorithm. A characteristic feature of simulated annealing is long computation times. Hence the possibility of executing the placement algorithm twice, once for ascertaining the final total interconnect length and the second for obtaining the final placement using the interconnect-area estimator, is unreasonable.

On the other hand, the random interconnection length can be measured in negligible computation time. A statistically meaningful number of random placements are examined and the average interconnection length is computed. Assume that it is possible to calculate the ratio ρ of the average interconnection length for random placement over optimized placement. Denoting L_R as the measured value of the random (total) interconnect length for a circuit, the final total interconnect length can be estimated by

$$L_o = \frac{L_R}{\rho} \tag{5.1}$$

The algorithms developed in this chapter enable the use of Eqn. 5.1 to compute an accurate estimate of the final total interconnect length for optimized placements.

The model for random placements yielded total interconnect lengths which were typically within five percent of the measured values. Furthermore, the model developed for optimized placements enabled the computation of values of ρ which were typically within seven percent of the measured ratios. The predicted results are obtained strictly from an analysis of the net list, with no *a priori* knowledge of the functionality of the circuit.

The organization of the remaining six sections of this chapter is now given. Section 5.2 presents the placement model which was used to develop the algorithms for estimating the average interconnection length. Previous approaches to the average interconnection length estimation problem are reviewed in Section 5.3. Section 5.4 introduces an algorithm for computing the average interconnection length for random placements, where it is assumed that all nets interconnect exactly two pins. Section 5.5 presents a model for the estimation of the average interconnection length for random cell placements in which nets may consist of any number of pins. An algorithm for estimating the average interconnection length for optimized placements is developed in Section 5.6. Finally, the chapter concludes with Section 5.7 which presents the results of comparisons of the models and algorithms for average interconnection length developed in this chapter with measured values obtained by using the placement algorithm of the previous chapter.

5.2 The Placement Model

The placement model used in this chapter assumes that N cells require placement. It is also assumed that each cell is square and the same size. Furthermore, the cells are to be placed on a $\sqrt{N} \times \sqrt{N}$ board, where $\sqrt{N}$ is assumed to be an integer. The area of each cell is given by the area of the board divided by N. This placement model, illustrated in Figure 5.1, will henceforth be referred to as the *checkerboard model.*

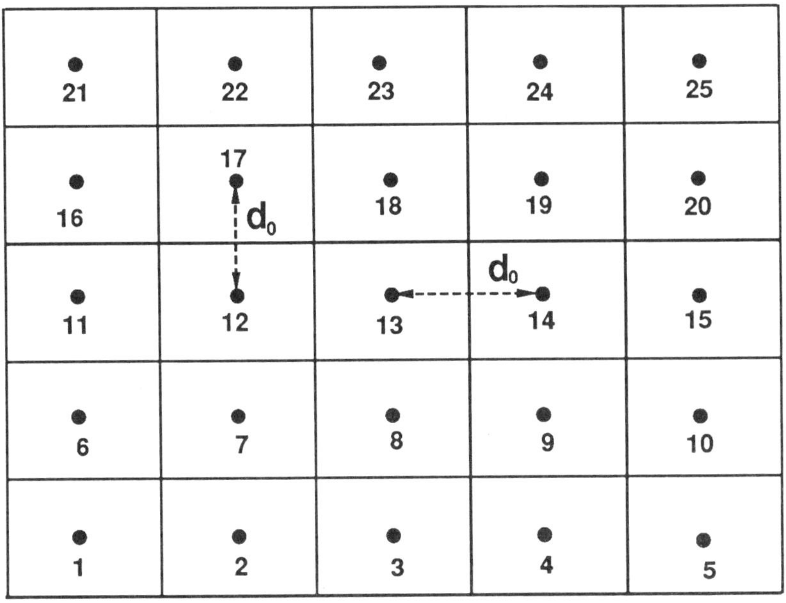

Figure 5.1 The checkerboard model. The center-to-center spacing of the cells is uniform.

In this model, the distance from the center of any cell c, where $c \in \{1, \cdots, 25\}$ in Figure 5.1, to the center of an adjacent cell is given by d_0. The value of d_0 is uniform throughout the array. Consequently, it is common to express interconnect lengths in terms of a multiple of d_0. That is, d_0 is normalized to a value of one.

A net is assumed to interconnect two or more cells. Initially, it will be assumed that the interconnection point on a cell is the center of the cell. The smallest enclosing rectangle of the interconnection points for a net defines the *net bounding box*. In Figure 5.2, a net interconnects the set of cells $\{7, 13, 14, 17, 23\}$. The net bounding box is as shown.

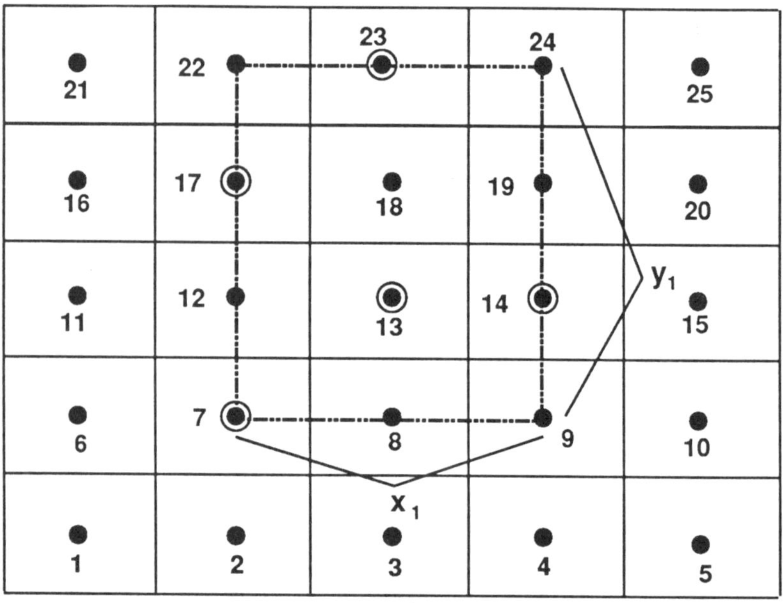

Figure 5.2 The net bounding box of a net interconnecting the set of cells {7, 13, 14, 17, 23} is shown.

The interconnection length for a net is assumed to be one half of the perimeter of the bounding box of the net.[1] This metric was selected since it is used by the placement algorithms of the two previous chapters. The x-span of the bounding box for net i, denoted by $x_s(i)$ is precisely the width of the bounding box for net i. Similarly, the y-span of net i, or $y_s(i)$, is equal to the height of the bounding box of net i. In Figure 5.2, the interconnect length for the indicated net is given by: $x_1 + y_1 = 2 + 3 = 5$. The total interconnect length L for a circuit of N_n nets can be expressed by:

$$L = \sum_{i=1}^{N_B} x_s(i) + y_s(i)$$

5.3 Previous Approaches

There have been previous attempts at *a priori* average interconnection length estimation. Gilbert derived expressions for the average minimum spanning tree lengths of randomly placed nets of n points.[2] Also, Donath developed a lower bound on average in-

[1] S. Goto and E. Kuh, "An Approach to the Two-Dimensional Placement Problem in Circuit Layout," (1978).

[2] E. N. Gilbert, "Random Minimal Trees," (1965).

terconnection length for random graphs.[1] Sutherland and Oestreicher developed a method for estimating routing track requirements for randomly placed printed circuit boards.[2] These approaches fail to account for placement optimization.

More recently, Donath developed an upper bound for the average interconnection length based on partitioning results.[3] This upper bound yielded significantly lower average interconnection lengths in comparison to bounds based on random placement. Donath showed that actual placements had average interconnection lengths which were about half the upper bound given by his theory. The approach is based on the experimentally observed terminal to cell ratio known as Rent's Rule.[4] On the highest level of hierarchy, a terminal is a pad. On other levels of hierarchy, a terminal is an interconnection point on the boundary of a super cell. The Rent Relationship is given by:

$$T = AC^p \tag{5.2}$$

where T is the average number of terminals required by a supercell containing an average of C cells, A is the average number of terminals per cell, and p, where $0 \le p \le 1$, is a constant for a given circuit.

The results derived by Donath were the following, where $\overline{R}$ is the average interconnection length for the checkerboard model with d_0 normalized to unity:

$$\begin{aligned} \overline{R} &\sim C^{p-1/2} , && p > 1/2 \\ \overline{R} &\sim \log C , && p = 1/2 \\ \overline{R} &\sim f(p) , && p < 1/2 \end{aligned} \tag{5.3}$$

where $f(p)$ is independent of C.

Donath stated that he had observed that the values of the *partitioning exponent* p ranged as high as 0.75 for highly parallel circuitry and as low as 0.47 for highly serialized circuitry. He showed that interconnect space is a very strong function of the character of the logic. Furthermore, he indicated that the $\overline{R}$ value for a random placement on a square array is given by:

$$\overline{R} = \frac{\sqrt{C}}{3} \tag{5.4}$$

This expression is actually incorrect. The correct expression is:

$$\overline{R} = \frac{2}{3}\sqrt{C} \tag{5.5}$$

and its derivation is the subject of Section 5.4.

While the upper bound expressed by Eqn. 5.3 is substantially more accurate for optimized placements than that given by Eqn. 5.5, the $\overline{R}$ result of Eqn. 5.3 cannot be used unless accurate knowledge of the circuit details is possessed. By its very nature, p is interpreted as the *degree of parallelism* of the logic circuit. For $p \ge 1/2$, apparently the usual case, the value of $\overline{R}$ is exponentially related to the value of p. Hence, accuracy in $\overline{R}$ can

---

[1] W. E. Donath, "Statistical Properties of the Placement of a Graph," (1968).

[2] I. Sutherland and D. Oestreicher, "How Big Should a Printed Circuit Board Be?," (1973).

[3] W. E. Donath, "Placement and Average Interconnect Lengths of Computer Logic," (1979).

[4] B. Landman and R. Russo, "On a Pin Versus Block Relationship for Partitions of Logic Graphs," (1971).

only be achieved if p is even more accurately known. This is not an easily quantifiable concept, so accuracy appears out of the question. Even when p is accurately known as a result of past experience, Donath points out that actual placements give $\overline{R}$ values of about half the upper bound expressed by Eqn. 5.3. The results derived in Sections 5.5 and 5.6 enable accurate estimations of the average interconnection length strictly from an analysis of the net list, with no *a priori* knowledge of the functionality of the circuit.

5.4 Average Interconnection Length for Random Placements under the Assumption of Two-Pin Nets

In this section, the average interconnection length for a net $\overline{R}_{r2}$ is derived for random cell placements in accordance with the checkerboard model. Further, each net is assumed to interconnect exactly two pins. Each possible cell location in the two-dimensional checkerboard model has coordinates given by (i, j) where $1 \le i, j \le \sqrt{N}$. Consider a net which connects to a cell c_1 at coordinates (i_1, j_1), which connects to a cell c_2 at coordinates (i_2, j_2). Permitting c_1 and c_2 to take on all possible locations on the checkerboard model and dividing by the number of possible combinations leads to the expression below for the average length of a net:

$$\overline{R}_{r2} = \frac{1}{N\,(N-1)} \sum_{i_1=1}^{\sqrt{N}} \sum_{j_1=1}^{\sqrt{N}} \sum_{i_2=1}^{\sqrt{N}} \sum_{j_2=1}^{\sqrt{N}} |i_1 - i_2| + |j_1 - j_2| \tag{5.6}$$

Simplification proceeds as follows:

$$\overline{R}_{r2} = \frac{1}{N\,(N-1)} \sum_{i_1=1}^{\sqrt{N}} \sum_{j_1=1}^{\sqrt{N}} \left[\sum_{i_2=1}^{\sqrt{N}} \sqrt{N}\,|i_1 - i_2| + \sum_{j_2=1}^{\sqrt{N}} \sqrt{N}\,|j_1 - j_2| \right] \tag{5.7}$$

$$\overline{R}_{r2} = \frac{\sqrt{N}}{N\,(N-1)} \sum_{i_1=1}^{\sqrt{N}} \sum_{j_1=1}^{\sqrt{N}} \left[\sum_{i_2=1}^{i_1} (i_1 - i_2) + \sum_{i_2=i_1+1}^{\sqrt{N}} (i_2 - i_1) + \sum_{j_2=1}^{j_1} (j_1 - j_2) + \sum_{j_2=j_1+1}^{\sqrt{N}} (j_2 - j_1) \right] \tag{5.8}$$

Further,

$$\overline{R}_{r2} = \frac{\sqrt{N}}{N\,(N-1)} \sum_{i_1=1}^{\sqrt{N}} \sum_{j_1=1}^{\sqrt{N}} \left[2\sum_{i_2=1}^{i_1} (i_1 - i_2) - \sum_{i_2=1}^{\sqrt{N}} (i_1 - i_2) + 2\sum_{j_2=1}^{j_1} (j_1 - j_2) - \sum_{j_2=1}^{\sqrt{N}} (j_2 - j_1) \right] \tag{5.9}$$

$$\overline{R}_{r2} = \frac{\sqrt{N}}{N(N-1)} \sum_{i_1=1}^{\sqrt{N}} \sum_{j_1=1}^{\sqrt{N}} \left[i_1^2 - (\sqrt{N}+1)\, i_1 + \frac{\sqrt{N}(\sqrt{N}+1)}{2} + j_1^2 - (\sqrt{N}+1)\, j_1 + \frac{\sqrt{N}(\sqrt{N}+1)}{2} \right] \qquad (5.10)$$

Continuing the simplification:

$$\overline{R}_{r2} = \frac{2}{(N-1)} \sum_{i_1=1}^{\sqrt{N}} i_1^2 - (\sqrt{N}+1)\, i_1 + \frac{\sqrt{N}(\sqrt{N}+1)}{2} \qquad (5.11)$$

Evaluating the summation for each of the three terms leads to:

$$\overline{R}_{r2} = \frac{\sqrt{N}(\sqrt{N}+1)}{(N-1)} \left[\frac{2\sqrt{N}+1}{3} - (\sqrt{N}+1) + \sqrt{N} \right] \qquad (5.12)$$

The final expression for the average interconnection length for random placements is then given by:

$$\overline{R}_{r2} = \frac{2}{3}\sqrt{N} \qquad (5.13)$$

5.4.1 Practical considerations

It is desired to compute the average interconnection length for random placements of industrial macro/custom cell circuits. Equation 5.13 is an expression for the average interconnection length in terms of the center-to-center cell spacing. Industrial macro/custom cell circuits generally have cells of various sizes. Hence it is necessary to compute an average center-to-center cell spacing. Given that the sum of the areas of all the cells is A_{tot}, the average cell area A_c is given by:

$$A_c = \frac{A_{tot}}{N} \qquad (5.14)$$

An average cell width W_c (or height) is then given by:

$$W_c = \sqrt{A_c} \qquad (5.15)$$

W_c represents the average center-to-center cell spacing for the checkerboard model.

Further, represent the total number of nets by n_t. The total interconnect length for a random placement of a macro/custom cell circuit is then given by:

$$\overline{L}_R = \frac{2}{3}\sqrt{N}\, n_t \sqrt{\frac{A_{tot}}{N}} \qquad (5.16)$$

5.5 Average Interconnection Length for Random Placements Having Nets of Arbitrary Pin Counts

The method developed here for computing the average interconnection length for random placements having nets of arbitrary pin counts is based on the following metric: The interconnection length for a net is assumed to be one half of the perimeter of its bounding box, defined in accordance with the checkerboard model presented in Section 5.2. In order to compare the theoretical results with the experimental, note that TimberWolfMC permits cells to overlap at the highest temperatures. In fact, at the initial temperature when the random total interconnect length is measured, every cell move is accepted. Hence, for the checkerboard model of *random placements*, it is necessary to permit more than one cell to occupy a checkerboard location.

Consider a *cell bounding box* whose width is m cells and whose height is s cells. The basic strategy is based on the following definition:

Definition 5.1:

A net n interconnecting a group of k cells, where the cells belong to the set $\{c_1, c_2, \cdots, c_k\}$, is said to *establish* a *cell bounding box* of size $m \times s$ under the following conditions: (1) $m = x_2 - x_1 + 1$. (2) $s = y_2 - y_1 + 1$. (3) There exists some $i_1 \in \{1, \cdots, k\}$ such that cell c_{i_1} is characterized by $x = x_1$. (4) There exists some $i_2 \in \{1, \cdots, k\}$ such that cell c_{i_2} is characterized by $x = x_2$. (5) There exists some $i_3 \in \{1, \cdots, k\}$ such that cell c_{i_3} is characterized by $y = y_1$. (6) There exists some $i_4 \in \{1, \cdots, k\}$ such that cell c_{i_4} is characterized by $y = y_2$.

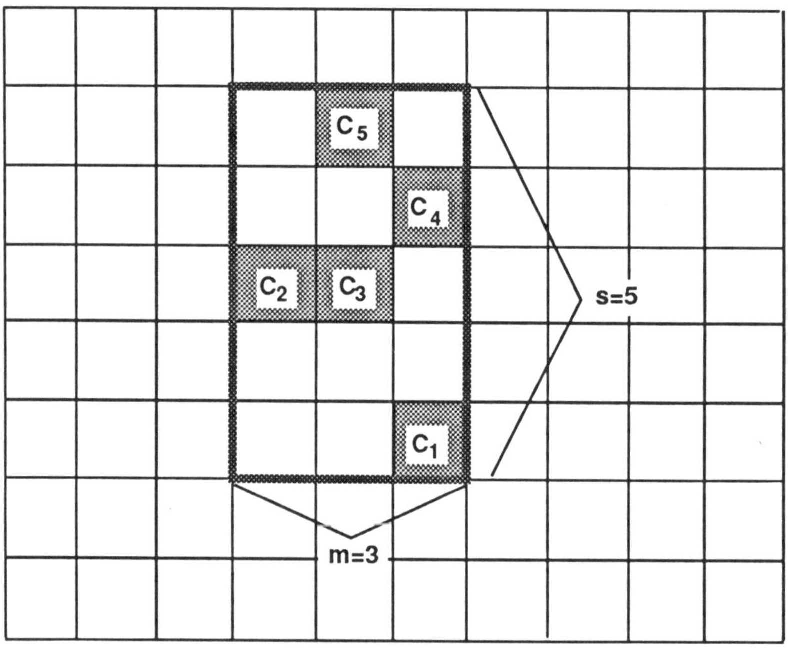

Figure 5.3 A net interconnecting cells C_1, C_2, C_3, C_4, and C_5 establishes the *cell bounding box* characterized by $m = 3$ and $s = 5$.

The *cell bounding box* for a net interconnecting cells C_1, C_2, C_3, C_4, and C_5 is shown in Figure 5.3. This net *establishes* the cell bounding box characterized by $m = 3$ and $s = 5$. Note that cell C_2 establishes the left boundary of the bounding box, cell C_1 establishes the bottom boundary of the bounding box, cells C_1 and C_4 establish the right boundary, and cell C_5 establishes the top boundary. Since it is assumed that all nets interconnect to the center of the cells, the interconnection length for this net is expressed by: $(m + s - 2)$.

One method of computing the average interconnection length for a net n proceeds as follows: (1) Enumerate all possible locations of the cells belonging to n. (2) For each configuration, find the cell bounding box and the associated values of m and s. (3) Average the quantity $(m + s - 2)$ over all configurations to yield the average interconnection length $\overline{R}_r$ for net n.

An easier method computes $\overline{R}_r$ for net n by essentially reversing the order of steps (1) and (2) above. That is, a cell bounding box characterized by an x-span given by m cells and a y-span given by s cells is *assumed.* Then, count all of the possible arrangements of the cells of n which *establish* this $m \times s$ cell bounding box. Denote this sum by $Q_{ms}(k)$. If the checkerboard model is characterized by a width M and a height S, the average interconnection length for a net n is given by:

$$\overline{R}_r = \frac{\sum_{m=1}^{M} \sum_{s=1}^{S} N_{ms}\ Q_{ms}(k)\ (m + s - 2)}{\sum_{m=1}^{M} \sum_{s=1}^{S} N_{ms}\ Q_{ms}(k)} \tag{5.17}$$

where N_{ms} is the number of positions on the checkerboard which can be occupied by the $m \times s$ bounding box. This quantity is given by:

$$N_{ms} = (M - m + 1)\ (S - s + 1) \tag{5.18}$$

Eqn. 5.17 can then be rewritten in the following manner:

$$\overline{R}_r = \frac{\sum_{m=1}^{M} \sum_{s=1}^{S} Q_{ms}(k)\ (M - m + 1)\ (S - s + 1)\ (m + s - 2)}{\sum_{m=1}^{M} \sum_{s=1}^{S} Q_{ms}(k)\ (M - m + 1)\ (S - s + 1)} \tag{5.19}$$

The computation of $Q_{ms}(k)$ remains. It is necessary to count all possible arrangements of the k cells of net n such that an $m \times s$ cell bounding box is established. Given a net consisting of k cells, there are seven possible *scenarios* which can give rise to an $m \times s$ bounding box. The value of $Q_{ms}(k)$ is then determined by the sum of the number of states yielded by each scenario. Note that some scenarios may not yield any states for the particular values of k, m and s.

In order to enumerate the number of states for each of the scenarios, it is necessary to special case the four corners of the checkerboard. This is due to the fact that cells in a corner establish *two* edges of the cell bounding box. On the other hand, cells on the left or right boundary columns, excluding the corners, establish only the respective single bounding box edge. Similarly, cells in the bottom or top rows of the bounding box, excluding the corners, establish only the corresponding single edge of the bounding box. In the remainder of this subsection, scenario number i for $0 \le i \le 4$ refers to the fact that exactly i of the bounding box corners are occupied by one or more of the k cells. Scenario number i for $5 \leq i \leq 7$ has no numerical significance.

Scenario 0:

There are no cells in any of the four corners of the $m \times s$ cell bounding box. In order to *establish* this bounding box, a net must have at least one cell in each of the leftmost and rightmost columns, and at least one cell in each of the bottommost and topmost rows of the bounding box. Note that for this scenario to exist for a given net, the bounding box must be characterized by $m \geq 3$ and $s \geq 3$. Furthermore, the net must consist of $k \geq 4$ cells. It is assumed that there are i_1 cells in the bottom row, i_2 cells in the top row, i_3 cells in the leftmost column, and i_4 cells in the rightmost column of the bounding box. The situation is depicted in Figure 5.4.

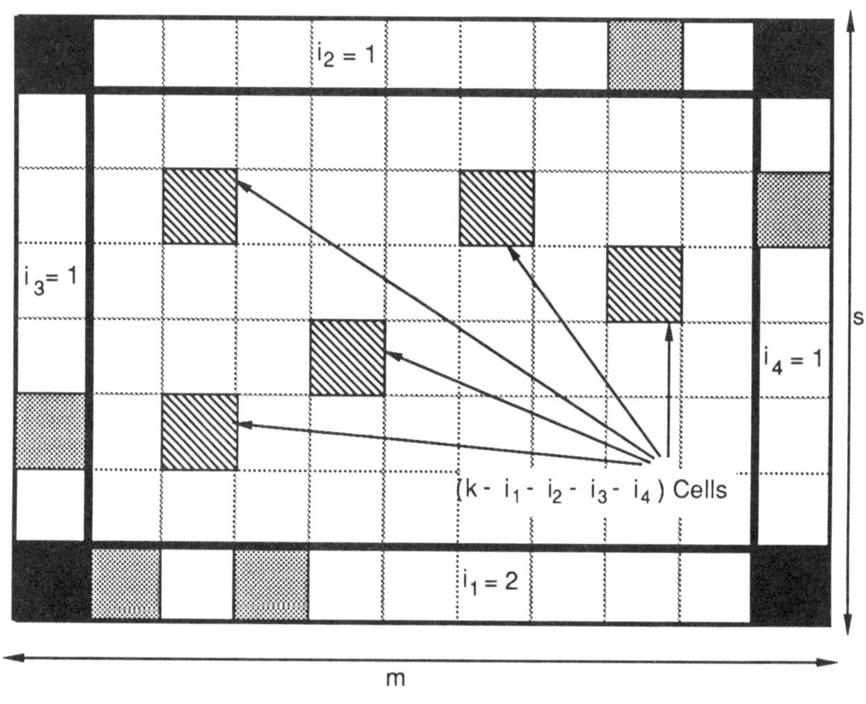

Figure 5.4 Scenario 0 for the random placement case: There are no cells in any of the four corners of the $m \times s$ bounding box. There are i_1 cells in the bottom row, i_2 cells in the top row, i_3 cells in the leftmost column, and i_4 cells in the rightmost column of the bounding box.

It follows that there are $(k-i_1-i_2-i_3-i_4)$ cells which do *not* occupy a bordering position, and are therefore said to occupy the *center region* of the bounding box. Since, as stated earlier, multiple cells are permitted to occupy a single board location, the number of possible arrangements in which there are i_1 cells in the bottom row, i_2 cells in the top row,

i_3 cells in the leftmost column, i_4 cells in the rightmost column, and $(k-i_1-i_2-i_3-i_4)$ cells in the center region, is given by:

$$\frac{k!}{i_1!\ i_2!\ i_3!\ i_4!\ (k-i_1-i_2-i_3-i_4)!} \tag{5.20}$$

The extent of the leftmost and rightmost columns is given by $(s-2)$ and the extent of the bottommost and topmost rows is given by $(m-2)$. The cells are assumed to be distinguishable (they have distinct names in practice), although multiple cells in one slot are assumed to be indistinguishable (to be consistent with the implementation of the placement algorithm). Of course, empty slots are indistinguishable.

Note that there are $(m-2)^{i_1}$ possible arrangements of the i_1 cells in the bottommost row. That is, each of the i_1 cells can be placed in any of the $(m-2)$ slots. Furthermore, there are: $(m-2)^{i_2}$ arrangements of the i_2 cells in the topmost row. Similar expressions can be derived for the leftmost and rightmost columns. Note that there are: $(ms-2m-2s+4)$ slots in the center region, and hence

$$(ms-2m-2s+4)^{(k-i_1-i_2-i_3-i_4)}$$

arrangements of cells in the center region.

A *state* will refer to an arrangement of the cells which establishes the $m \times s$ bounding box. The overall expression for the total number of states, denoted as σ_0, is obtained by summing over all possible values of i_1, i_2, i_3, and i_4. Note that $\sigma_0 = 0$ unless $m \geq 3$, $s \geq 3$, and $k \geq 4$, in which case σ_0 is given by:

$$\sigma_0 = \sum_{i_1=1}^{k-3} \sum_{i_2=1}^{k-2-i_1} \sum_{i_3=1}^{k-1-i_1-i_2} \sum_{i_4=1}^{k-i_1-i_2-i_3} \left[\frac{k!\ (m-2)^{i_1}\ (m-2)^{i_2}}{i_1!\ i_2!\ i_3!\ i_4!} \cdot \frac{(s-2)^{i_3}\ (s-2)^{i_4}\ (ms-2m-2s+4)^{(k-i_1-i_2-i_3-i_4)}}{(k-i_1-i_2-i_3-i_4)!} \right] \tag{5.21}$$

Scenario 1:

There are cells in only one of the four corners of the $m \times s$ cell bounding box. In order to *establish* this bounding box, a net must have at least one cell in one of the corners and at least one cell in the nonadjacent outermost row and column. The situation is depicted in Figure 5.5.

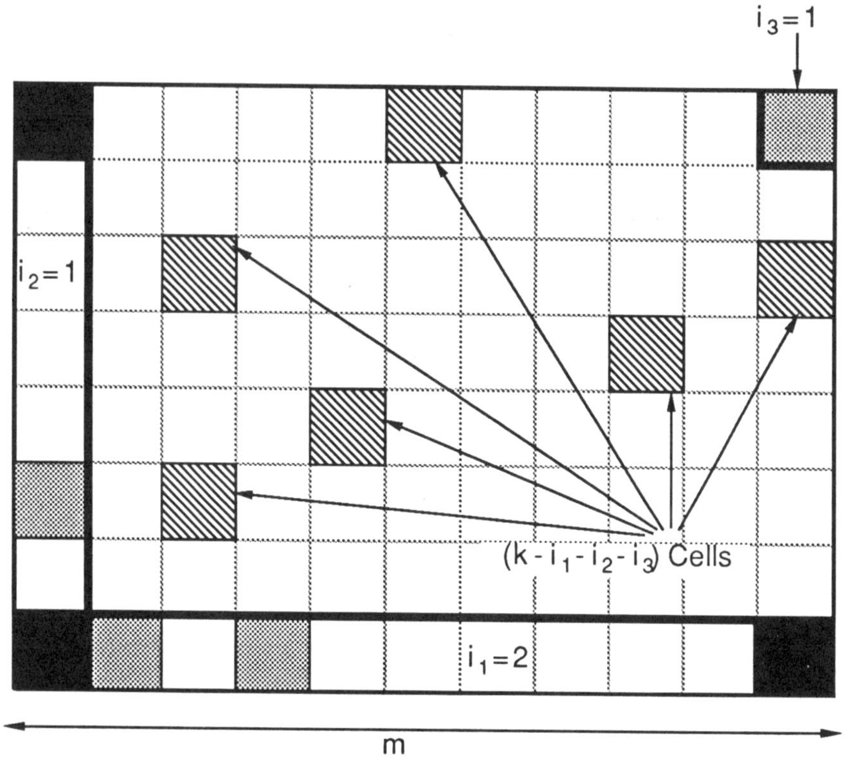

Figure 5.5 Scenario 1 for the random placement case: There are cells in only one of the four corners of the $m \times s$ bounding box. There are i_1 cells in one of the outer rows, i_2 cells in the adjacent outer column, and i_3 cells in the opposite corner of the bounding box.

Four permutations arise, one generated for each of the different corners which could be the *occupied* corner. Note that for this scenario to exist for a given net, the bounding box must be characterized by $m \geq 3$ and $s \geq 3$. Furthermore, the net must consist of $k \geq 3$ cells. With respect to Figure 5.5, it is assumed that there are i_1 cells in the bottom row. There are i_2 cells in the leftmost column, and i_3 cells in the upper right corner.

The expression for the total number of states is denoted by σ_1. This expression was developed using arguments which closely parallel those presented for the case of scenario 0. Note that $\sigma_1 = 0$ unless $m \geq 3$, $s \geq 3$, and $k \geq 3$, in which case σ_1 is given by:

$$\sigma_1 = \sum_{i_1=1}^{k-2} \sum_{i_2=1}^{k-1-i_1} \sum_{i_3=1}^{k-i_1-i_2} \frac{4k!\ (m-2)^{i_1}\ (s-2)^{i_2}\ (ms-m-s)^{(k-i_1-i_2-i_3)}}{i_1!\ i_2!\ i_3!\ i_4!\ (k-i_1-i_2-i_3)!} \tag{5.22}$$

Scenario 2A:

There are cells in exactly two opposite corners of the $m \times s$ cell bounding box. In order to *establish* this bounding box, a net must have at least one cell in each of the two corners. The situation is depicted in Figure 5.6.

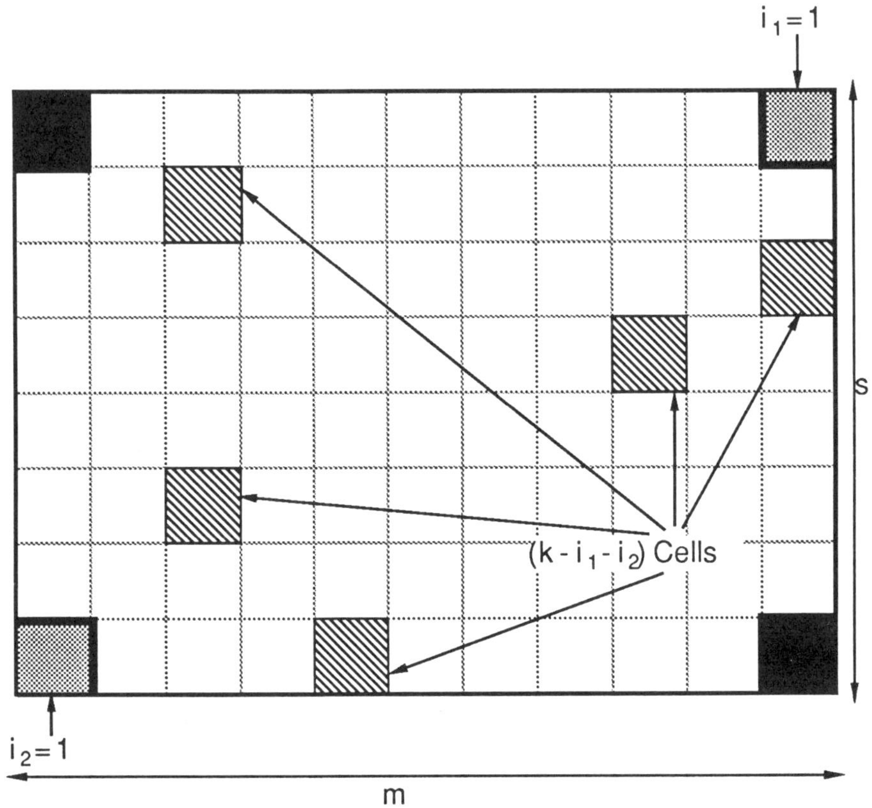

Figure 5.6 Scenario 2A for the random placement case: There are cells in exactly two opposite corners of the $m \times s$ bounding box. There are i_1 cells in one of the corners and i_2 cells in the other corner.

Two permutations arise, one generated for each of the two sets of opposite-corner pairs which are occupied. Note that for this scenario to exist for a given net, the bounding box must be characterized by $m \geq 2$ and $s \geq 2$. Furthermore, the net must consist of $k \geq 2$ cells. With respect to Figure 5.6, it is assumed that there are i_1 cells in one of the corners and i_2 cells in the other corner.

The expression for the total number of states is denoted by σ_{2A}. Note that $\sigma_{2A} = 0$ unless $m \geq 2$, $s \geq 2$, and $k \geq 2$, in which case σ_{2A} is given by:

$$\sigma_{2A} = \sum_{i_1=1}^{k-1} \sum_{i_2=1}^{k-i_1} \frac{2k!\ (ms-4)^{(k-i_1-i_2)}}{i_1!\ i_2!\ (k-i_1-i_2)!} \tag{5.23}$$

Scenarios 2B and 2C:

There are cells in exactly two adjacent corners of the $m \times s$ cell bounding box. In order to *establish* this bounding box, a net must have at least one cell in each of the two corners. The situation is depicted in Figure 5.7.

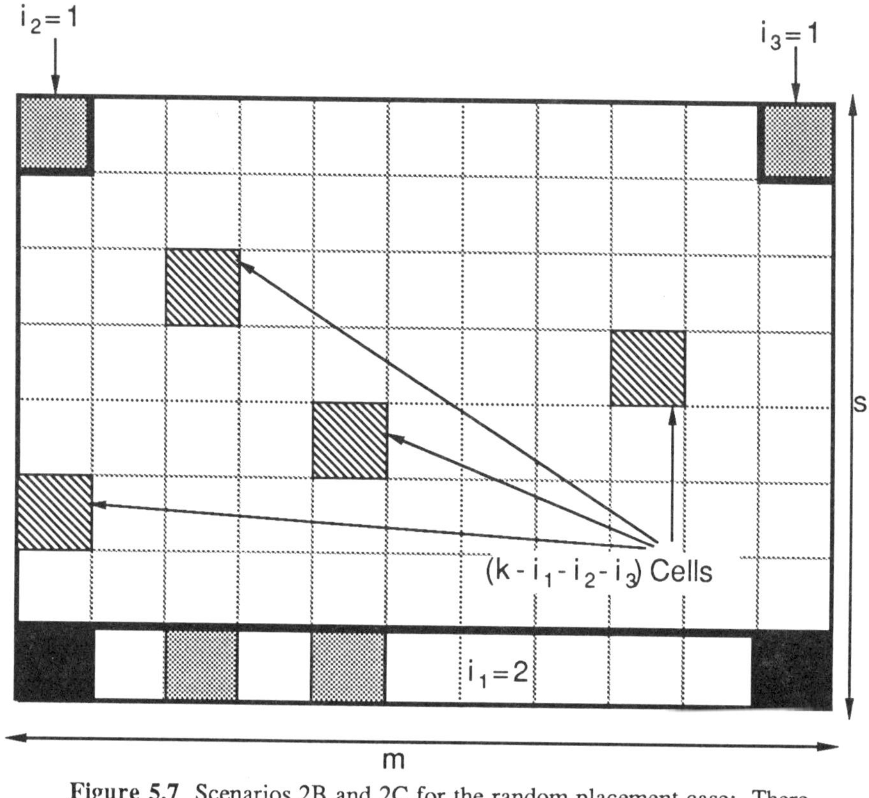

Figure 5.7 Scenarios 2B and 2C for the random placement case: There are cells in exactly two adjacent corners of the $m \times s$ bounding box. There are i_2 cells in one of the corners and i_3 cells in the other corner. Also, there are i_1 cells in the row or column bordered by the two unoccupied corners.

Scenario 2B denotes the situation when the two occupied corners lie on the same outer row, as depicted in Figure 5.7. For this case, two permutations arise, one in which the occupied corners are in the topmost row and one in which they are in the bottommost row. Note that for this scenario to exist for a given net, the bounding box must be characterized by $m \geq 3$ and $s \geq 2$. Furthermore, the net must consist of $k \geq 3$ cells. With respect to Figure 5.7, it is assumed that there are i_1 cells in one of the outermost rows, i_2 cells in one of the corners and i_3 cells in the other occupied corner.

The expression for the total number of states is denoted by σ_{2B}. Note that $\sigma_{2B} = 0$ unless $m \geq 3$, $s \geq 2$, and $k \geq 3$, in which case σ_{2B} is given by:

$$\sigma_{2B} = \sum_{i_1=1}^{k-2} \sum_{i_2=1}^{k-1-i_1} \sum_{i_3=1}^{k-i_1-i_2} \frac{2k!\ (m-2)^{i_1}\ (ms-m-2)^{(k-i_1-i_2-i_3)}}{i_1!\ i_2!\ i_3!\ (k-i_1-i_2-i_3)!} \tag{5.24}$$

Scenario 2C denotes the situation when the two occupied corners lie on the same outer column. For this scenario to exist for a given net, the bounding box must be characterized by $m \geq 2$ and $s \geq 3$. Furthermore, the net must consist of $k \geq 3$ cells. That is, $\sigma_{2C} = 0$ unless $m \geq 2$, $s \geq 3$, and $k \geq 3$, in which case σ_{2C} is given by:

$$\sigma_{2C} = \sum_{i_1=1}^{k-2} \sum_{i_2=1}^{k-1-i_1} \sum_{i_3=1}^{k-i_1-i_2} \frac{2k!\ (s-2)^{i_1}\ (ms-s-2)^{(k-i_1-i_2-i_3)}}{i_1!\ i_2!\ i_3!\ (k-i_1-i_2-i_3)!} \tag{5.25}$$

Scenario 3:

For a net to establish this bounding box, there must be at least one cell in exactly three of the corners. The situation is depicted in Figure 5.8.

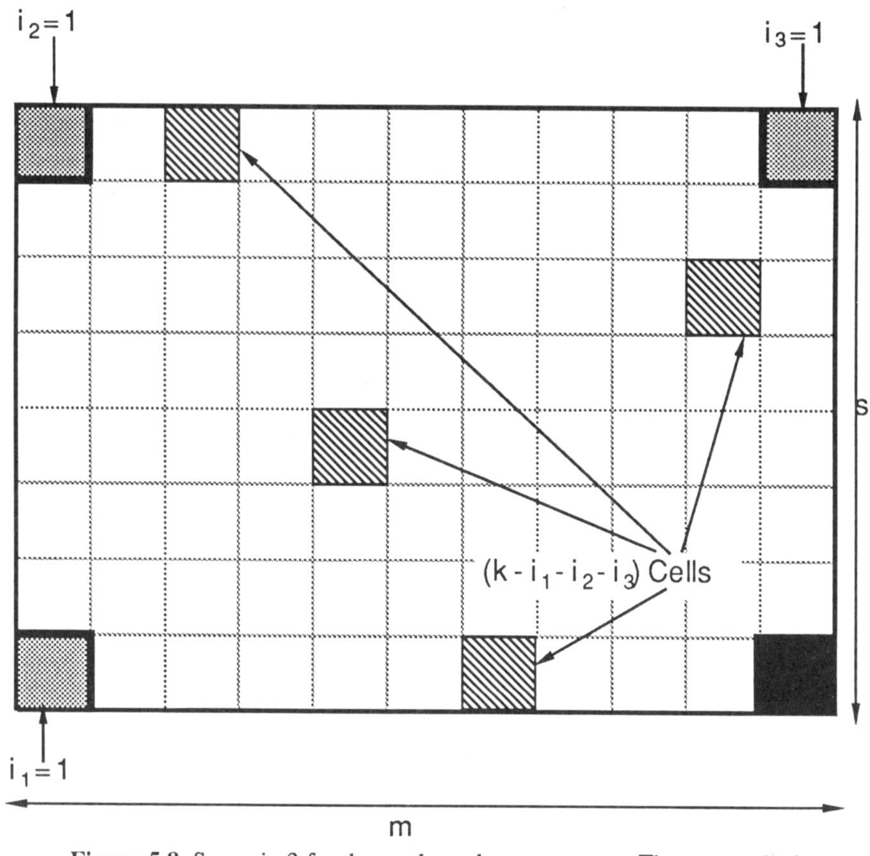

Figure 5.8 Scenario 3 for the random placement case: There are cells in exactly three of the four corners of the $m \times s$ bounding box. There are i_1, i_2, and i_3 cells in each of the three corners, respectively.

There are i_1, i_2, and i_3 cells in each of the three corners, respectively. For this scenario to exist for a given net, the bounding box must be characterized by $m \geq 2$ and $s \geq 2$. Furthermore, the net must consist of $k \geq 3$ cells. That is, $\sigma_3 = 0$ unless $m \geq 2$, $s \geq 2$, and $k \geq 3$, in which case σ_3 is given by:

$$\sigma_3 = \sum_{i_1=1}^{k-2} \sum_{i_2=1}^{k-1-i_1} \sum_{i_3=1}^{k-i_1-i_2} \frac{4k!\,(ms-4)^{(k-i_1-i_2-i_3)}}{i_1!\,i_2!\,i_3!\,(k-i_1-i_2-i_3)!} \tag{5.26}$$

Scenario 4:

For a net to establish this bounding box, there must be at least one cell in each of the four corners. The situation is depicted in Figure 5.9.

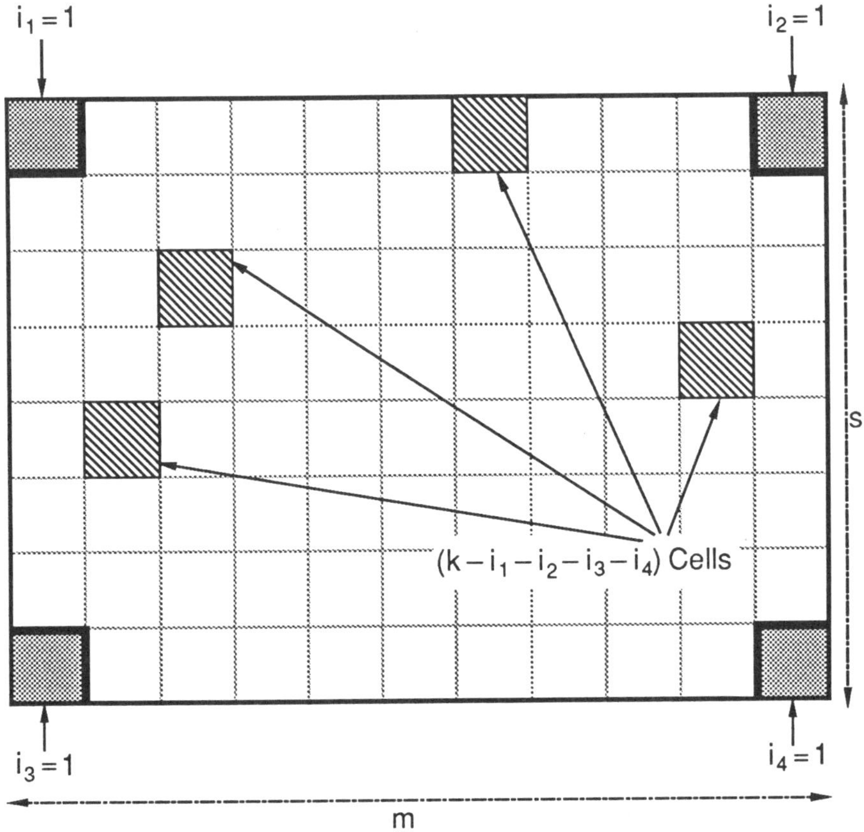

Figure 5.9 Scenario 4 for the random placement case: There are cells in each of the four corners of the $m \times s$ bounding box. There are i_1, i_2, i_3, and i_4 cells in each of the four corners, respectively.

There are i_1, i_2, i_3, and i_4 cells in each of the four corners, respectively. For this scenario to exist for a given net, the bounding box must be characterized by $m \geq 2$ and $s \geq 2$. Furthermore, the net must consist of $k \geq 4$ cells. That is, $\sigma_4 = 0$ unless $m \geq 2$, $s \geq 2$, and $k \geq 4$, in which case σ_4 is given by:

$$\sigma_4 = \sum_{i_1=1}^{k-2} \sum_{i_2=1}^{k-2-i_1} \sum_{i_3=1}^{k-1-i_1-i_2} \sum_{i_4=1}^{k-i_1-i_2-i_3} \frac{k!\,(ms-4)^{(k-i_1-i_2-i_3-i_4)}}{i_1!\,i_2!\,i_3!\,i_4!\,(k-i_1-i_2-i_3-i_4)!} \qquad (5.27)$$

Scenario 5:

Scenarios 5, 6, and 7 represent the bounding boxes which are actually not rectangles, but rather rows, columns, or single squares, respectively, in the checkerboard model. For a net to establish a scenario 5 bounding box, the cells belonging to the net must all lie in one row of the checkerboard model. There must be at least one cell at each endpoint of the $m \times 1$ bounding box, or row. The situation is illustrated in Figure 5.10.

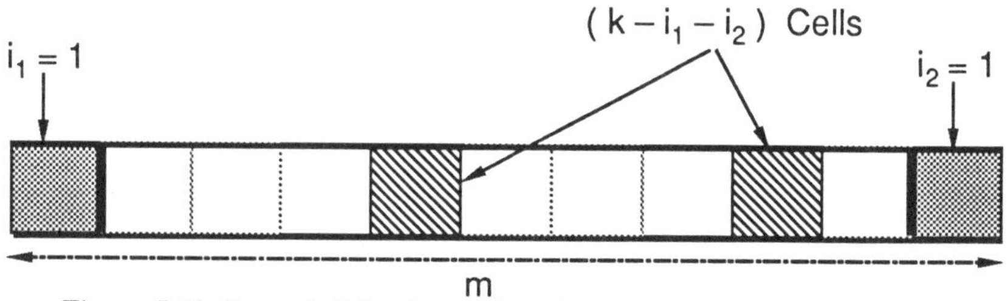

Figure 5.10 Scenario 5 for the random placement case: There are cells in both of the two end points of the $m \times 1$ bounding box. There are i_1 cells at one end and i_2 cells at the other end.

There are i_1 cells at one end and i_2 cells at the other end. For this scenario to exist for a given net, the bounding box must be characterized by $m \geq 2$ and $s = 1$. Furthermore, the net must consist of $k \geq 2$ cells. Note that $\sigma_5 = 0$ unless $s = 1$, $m \geq 2$ and $k \geq 2$, in which case σ_5 is given by:

$$\sigma_5 = \sum_{i_1=1}^{k-1} \sum_{i_2=1}^{k-i_1} \frac{k!\,(m-2)^{(k-i_1-i_2)}}{i_1!\,i_2!\,(k-i_1-i_2)!} \tag{5.28}$$

Scenario 6:

For a net to establish a scenario 6 bounding box, the cells belonging to the net must all lie in one column of the checkerboard model. There must be $i_1 \geq 1$ cells at one endpoint and $i_2 \geq 1$ cells at the other endpoint of the $1 \times s$ bounding box, or column. For this scenario to exist for a given net, the bounding box must be characterized by $m = 1$ and $s \geq 2$. Furthermore, the net must consist of $k \geq 2$ cells. Note that $\sigma_6 = 0$ unless $m = 1$, $s \geq 2$ and $k \geq 2$, in which case σ_6 is given by:

$$\sigma_6 = \sum_{i_1=1}^{k-1} \sum_{i_2=1}^{k-i_1} \frac{k!\,(s-2)^{(k-i_1-i_2)}}{i_1!\,i_2!\,(k-i_1-i_2)!} \tag{5.29}$$

Scenario 7:

For a net to establish a scenario 7 bounding box, *all* of the cells must lie in a single square in the checkerboard model. Note that $\sigma_7 = 0$ unless $m = 1$ and $s = 1$, in which case σ_7 is given by:

$$\sigma_7 = 1 \tag{5.30}$$

The expression for the total number of states $Q_{ms}(k)$ for given values of k, m, and s are summarized in Table 5.1. Note that for a given set of values of k, m, and s, only one row in the table is appropriate.

Case	k	m	s	The Total Number of States $Q_{ms}(k)$:
1	≥4	≥3	≥3	$\sigma_0 + \sigma_1 + \sigma_{2A} + \sigma_{2B} + \sigma_{2C} + \sigma_3 + \sigma_4$
2	≥4	≥3	2	$\sigma_{2A} + \sigma_{2B} + \sigma_3 + \sigma_4$
3	≥4	2	≥3	$\sigma_{2A} + \sigma_{2C} + \sigma_3 + \sigma_4$
4	≥4	2	2	$\sigma_{2A} + \sigma_3 + \sigma_4$
5	≥2	>1	1	σ_5
6	≥2	1	>1	σ_6
7	≥2	1	1	1
8	3	≥3	≥3	$\sigma_1 + \sigma_{2A} + \sigma_{2B} + \sigma_{2C} + \sigma_3$
9	3	≥3	2	$\sigma_{2A} + \sigma_{2B} + \sigma_3$
10	3	2	≥3	$\sigma_{2A} + \sigma_{2C} + \sigma_3$
11	3	2	2	$\sigma_{2A} + \sigma_3$
12	2	≥2	≥2	σ_{2A}

Table 5.1 The random placement problem. Twelve cases for the computation of $Q_{ms}(k)$ for nets of k cells to establish an $m \times s$ bounding box.

The random interconnection length $\overline{R}_r$ is obtained by using Eqn. 5.19 (repeated here) where the appropriate expression for $Q_{ms}(k)$ from Table 5.1 is inserted.

$$\overline{R}_r = \frac{\sum_{m=1}^{M} \sum_{s=1}^{S} Q_{ms}(k)\ (M - m + 1)\,(S - s + 1)\,(m + s - 2)}{\sum_{m=1}^{M} \sum_{s=1}^{S} Q_{ms}(k)\ (M - m + 1)\ (S - s + 1)} \tag{5.31}$$

If nets of various numbers of cells are present in a circuit, the usual case, then Eqn. 5.31 must be applied for each net size. The overall expression for $\overline{R}_r$ is the following:

$$\overline{R}_r = \frac{\sum_{k=2}^{k_{max}} \sum_{m=1}^{M} \sum_{s=1}^{S} N(k)\ Q_{ms}(k)\ (M - m + 1)\,(S - s + 1)\,(m + s - 2)}{\sum_{k=2}^{k_{max}} \sum_{m=1}^{M} \sum_{s=1}^{S} N(k)\ Q_{ms}(k)\ (M - m + 1)\ (S - s + 1)} \tag{5.32}$$

where $N(k)$ is the number of nets of size k cells.

5.5.1 Results

The model for the average interconnection length of random placements with multi-pin nets was tested on 9 industrial macro/custom cell circuits. The characteristics of the 9 circuits are presented in Table 5.2.

Circuit	No. Cells	No. Nets	No. Pins
i1	33	82	452
p1	11	44	309
x1	10	183	762
i2	23	109	577
i3	18	33	102
l1	62	551	4309
d2	20	586	1776
d1	17	288	837
d3	17	121	665

Table 5.2 Characteristics of the nine industrial macro/custom cell circuits.

Table 5.3 gives the net distributions *N(k)* with respect to the number of cells per net *k* for the 9 circuits.

	N(k): The Number of Nets With *k* Cells:														
Ckt	k=2	3	4	5	6	7	8	9	10	11	12	13	14	15	≥16
i1	65	11	1			1									
p1	14	12	6	6	4	2									
x1	128	44	5		1	1	1		3						
i2	86	12	5	1	5										
i3	13	17	3												
l1	307	93	48	15	29	4	4	2	6	11	2	8	13	3	6
d2	477	66	12	14	10	3	4								
d1	169	70	4	2	38	5									
d3	29	16	12	17	19	8	7	9							

Table 5.3 Net distributions with respect to the number of cells for the nine industrial macro/custom cell circuits.

Note that Eqn. 5.32 coupled with Table 5.1 gives the *precise* value of $\overline{R}_r$ for circuits with multipin nets which obey the checkerboard model. *Island-style* gate arrays, presented in the Appendix, obey this model. Hence, for these circuits, the experimental measurements of the average interconnection length for random placements *must* coincide with the so-called predicted value given by Eqn. 5.32 and Table 5.1. If not, the error lies with the measurements and not with the theoretical model.

The nine industrial circuits presented in Table 5.2 certainly do not obey the checkerboard model since the circuit image is not a rectangular array, and furthermore, the cells sizes and shapes vary considerably. This can be seen from Table 5.4 which indicates the standard deviation of the cell areas (σ_A) divided by the average cell area ($\overline{A}_c$). Note that circuits i2, l1, and d2 have especially wide variances in cell area, as does circuit d3.

Circuit	No. Cells	$\frac{\sigma_A}{\overline{A}_c}$
i1	33	2.91
p1	11	2.6
x1	10	1.64
i2	23	7.55
i3	18	1.72
l1	62	8.12
d2	20	8.55
d1	17	1.95
d3	17	3.78

Table 5.4 Ratio of the standard deviation of the average cell area (σ_A) over the average cell area ($\overline{A}_c$) for the nine circuits.

The results given by Eqn. 5.32 and Table 5.1 were nonetheless applied to the nine circuits. For each of the nine circuits, the pin positions on the cells were dispersed along their edges. Since all cell orientations were visited during the measurement process, the average pin position was located at the center of the cell. Each cell was assumed to have the same size, and the circuit image was assumed to be an $M \times S$ array. The sum of the cell areas divided by the number of cells N_C yielded the average cell area, represented by $\overline{A}_c$. The size of the rectangular image array was determined by first computing $M = S = \sqrt{N_C}$, and then adding one to S if the new product of $M \cdot S$ did not exceed N_C. The choice of making the circuit image as square as possible was done to match the target aspect ratio given to TimberWolfMC, used to obtain the measurement data. Table 5.5 shows the values of M and S, as well as the square root of $\overline{A}_c$, which can be interpreted as the average cell side.

Circuit	Cells	Nets	M	S	$\overline{R}_r$	$\overline{R}_{r2}$	$\sqrt{\overline{A}_c}$
i1	33	82	5	6	3.89	3.83	267
p1	11	44	3	3	3.26	2.21	640
x1	10	183	3	3	2.07	2.11	1987
i2	23	109	4	5	3.28	3.20	362
i3	18	33	4	4	3.32	2.83	740
l1	62	551	7	8	6.63	5.25	980
d2	20	586	4	5	3.22	2.98	2367
d1	17	288	4	4	3.26	2.75	195
d3	17	121	4	4	4.30	2.75	7110

Table 5.5 Predicted $\overline{R}_r$ and $\overline{R}_{r2}$ as well as the square root of the average cell area for the nine industrial macro/custom cell circuits. The $M \times S$ checkerboard model dimensions are given for each circuit.

Table 5.5 also indicates the value of $\overline{R}_r$ as computed by Eqn. 5.32 and Table 5.1, and compares it with the two-pin-net model ($\overline{R}_{r2} = \frac{2}{3}\sqrt{N_c}$) developed in Section 5.4. Note that the two-pin-net model deviates substantially for circuits p1, l1, and d3. Furthermore, for circuits i3, d2, and d1, the two-pin-net model shows moderate deviation.

The comparison of the experimentally measured *total* interconnect length versus the results produced by the multipin net model of Eqn. 5.32 is shown in Table 5.6.

Ckt	Cells	Nets	Pins	Predicted Random Wire Length ($\times 10^6$)	Measured Random Wire Length ($\times 10^6$)	Deviation (percent)
i1	33	82	452	85.1	78.5	7.7
p1	11	44	309	91.8	100	8.9
x1	10	183	762	753	770	2.2
i2	23	109	577	129	135	4.6
i3	18	33	102	8.1	7.8	3.7
l1	62	551	4309	3580	3663	2.3
d2	20	586	1776	4453	4181	6.1
d1	17	288	837	183	166	9.2
d3	17	121	665	3699	3451	6.7
avg.						5.6

Table 5.6 Comparison of predicted and measured total interconnect length for random placement.

Note that the measured total interconnect length values are typically within 5 percent of the predicted values. The computational complexity is dominated by the computation of the σ values. Note that the computation of σ_0 is actually $O(k^4)$. However, from Table 5.3, the values of k are typically not much larger than 10. Even much larger circuits, in terms of the number of cells and nets, would not be expected to have larger values of k.

5.6 A Model for Optimized Placement

The previous subsection developed an expression for the average interconnection length for the case of random placements of macro/custom cell circuits. In this subsection, a notion of an optimized placement will be described. This concept will then be used to develop an algorithm for computing the average interconnection length of a net for optimized placements.

The development of the model for optimized placement is also based on the checkerboard model. The key notion is that, given a cell c_0, a placement optimization algorithm will seek to place all the cells connected to c_0, represented by the set C, *close* to c_0. Before clarifying the notion of *close*, a method of computing the average number of cells connected to a cell is introduced.

5.6.1 The average number of other cells connected to a cell

Define $\overline{n}_c$ as the average number of nets connected to a cell. Further, define $\overline{c}_n$ as the average number of cells per net. Note that it is not sufficient to compute $\overline{C}$ by taking the product of the average number of nets connected to a cell and one less than the average number of cells per net. That is, the expression

$$\overline{C} = \overline{n}_c(\overline{c}_n - 1) \tag{5.33}$$

fails to properly account for nets which interconnect the same set of cells. For example, consider the simple case of the placement of a circuit consisting of two cells and ten nets. Then the product of the number of nets per cell (10) and the number of cells per net (2) minus one is ten. This suggests that a cell connects to ten other cells which is not possible since there are only two cells in the circuit.

Define k_i as the number of *distinct* cells connected to cell i and let N_c represent the total number of cells. Another possible expression $\overline{C}$ is the following:

$$\overline{C} = \frac{\sum_{i=1}^{N_c} k_i}{N_c} \tag{5.34}$$

Note that this method completely fails if the circuit has a net which interconnects the set of all cells, for example, a power, ground, or possibly a clock line. That is, in this case, if just one net interconnects the set of all cells, then $k = N_c$, and therefore $\overline{C} = N_c - 1$, independent of the other nets. One could contemplate not including such *large* nets in the calculation. Drawing the line as to just how big a net must be in order to classify it as too big, unfortunately, was found to be an impossible task for actual industrial circuits. Too often, there was almost a continuum of net sizes. Also, for circuits containing fewer than 10 cells or so, it truly is impossible to draw the line. Since TimberWolfMC is used quite often for circuits of this size, a new method of computing C was required.

5.6.1.1 The new method

The method introduced here avoids the above pitfalls. Two key parameters from the net list are used to compute C. The first is n_c^d, the average number of *distinct* nets per cell. The list of cells interconnected by each net in the net list is noted. That is, each net is represented by a list of cells. The number of *distinct* lists is defined to be the total number of distinct nets for the circuit. This number divided by the number of cells yields n_c^d.

The second parameter extracted from the net list is c_d, the average number of other cells per *distinct* net. This is computed directly from the distinct lists of cells, mentioned above.

A recurrence relation is developed to compute C. First, assume that n_c^d and c_d are integral. Further, let the total number of cells in the circuit be represented by N_c. Conceptually, computing C is equivalent to drawing c_d objects at a time out of a pool of N_c objects (which are replenished after each draw). After n_c^d draws, the number of distinct objects

Let ζ represent the number of distinct objects (cells) drawn so far. Then, one additional draw will yield a total of $(\zeta + i)$ distinct objects $(0 \leq i \leq c_d)$ with (unnormalized) probability P_i^ζ:

$$P_i^\zeta = \binom{\zeta}{c_d - i}\binom{N_c - \zeta}{i}$$

$$= \frac{\zeta!}{(c_d - i)!\,(\zeta - (c_d - i))!}\;\frac{(N_c - \zeta)!}{i!\,(N_c - \zeta - i)!} \tag{5.35}$$

That is, P_i^ζ is given by the product of (1) the number of ways that i new objects can be drawn from the $(N_c - \zeta)$ objects not yet drawn and (2) the number of ways that $(c_d - i)$ *old* objects can be drawn from those objects *already* drawn.

Figure 5.11 illustrates the *probability* tree which results from repeated applications of Eqn. 5.35, starting from the root which represents the c_d distinct objects (cells) obtained after the first draw.

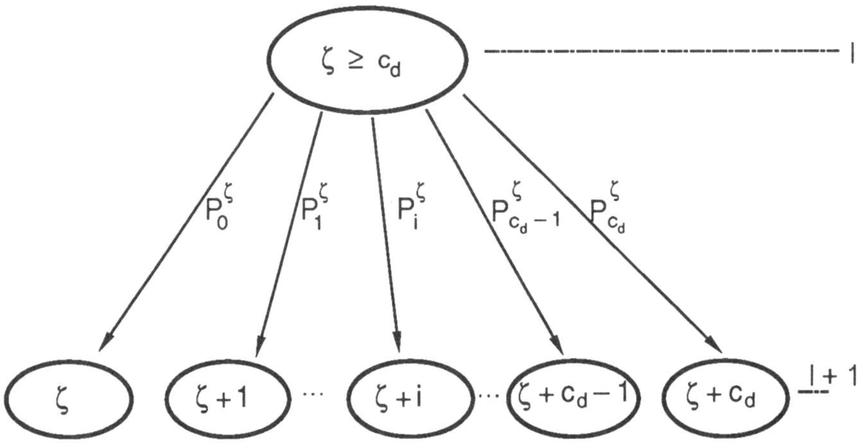

Figure 5.11 The probability tree which results from repeated applications of Eqn. 5.35, starting from the root which represents the c_d distinct cells obtained after the first draw. Each level l in the tree, where level 1 constitutes the root, represents the l-th draw. The number of distinct objects obtained by following a path in the tree from the root to a particular node is represented by the quantity inside the oval-shaped node. The probability affixed to an edge in the tree is given by Eqn. 5.35, where ζ is the number of distinct objects at the parent node and $(\zeta + i)$ is the number of distinct objects at the child node.

Each level l in the tree, where level 1 constitutes the root, represents the l-th draw. The circle depicting each node contains a value ζ which represents the number of distinct objects obtained by following the path in the tree which arrives at this node. The probability affixed to an edge in the tree is given by Eqn. 5.35, where ζ is the number of distinct objects at the parent node and $(\zeta + i)$ is the number of distinct objects at the child node. The probability of arriving at a particular node after the *l-th* draw is given by a product of $(l - 1)$ probabilities of the form expressedy by Eqn. 5.35.

The expected value of the number of distinct objects drawn (C) for $l = n_c^d$ draws is computed by completing the tree for n_c^d levels using Eqn. 5.35. For each leaf node, first compute the probability of arriving at the node (a product of $l - 1$ probabilities) and then multiply by the number of distinct objects represented by the leaf node. The sum of this quantity for all leaf nodes yields, after normalization, the expected value of C:

$$\overline{C} = \frac{\sum_{l=1}^{\#\text{leaves}} \left[\prod_{\lambda=2}^{n_c^d} P_{i_\lambda}^{\zeta_\lambda} \right] \zeta_l}{\sum_{l=1}^{\#\text{leaves}} \left[\prod_{\lambda=2}^{n_c^d} P_{i_\lambda}^{\zeta_\lambda} \right]} \tag{5.36}$$

where ζ_l represents the value of ζ for a leaf node at level l.

5.6.1.2 Practical considerations

In practice, when analyzing a net list, the values of $l \equiv n_c^d$ and $h \equiv c_d$ are usually both nonintegral. The expected value of C is computed by linear interpolation. First, consider only the floor of l. Now take the floor of h and use Eqn. 5.36 to compute C_{h1}. Next, take the ceiling of h and compute C_{h2}. Perform a linear interpolation (based on the actual value of h) of C_{h1} and C_{h2} to find C_1.

Second, consider only the ceiling of l. Take the floor of h and use Eqn. 5.36 to compute C_{h3}. Next, take the ceiling of h and compute C_{h4}. Perform a linear interpolation (again, based on the actual value of h) of C_{h3} and C_{h4} to find C_2. Finally, the expected value of C (that is, $\overline{C}$) is computed by linearly interpolating C_1 and C_2, based on the actual value of l.

5.6.1.3 Results

Table 5.7 presents the characteristics of the nine industrial circuits in terms of the number of distinct nets per cell and the number of other cells per distinct net.

Circuit	Cells	Nets	Pins	n_c^d	c_d
i1	33	82	452	1.45	2.96
p1	11	44	309	1.45	2.50
x1	10	183	762	4.70	2.02
i2	23	109	577	2.35	1.72
i3	18	33	102	1.67	1.73
l1	62	551	4309	3.74	3.96
d2	20	586	1776	5.70	2.16
d1	17	288	837	3.71	1.86
d3	17	121	665	2.71	3.80

Table 5.7 Characteristics of the nine industrial macro/custom cell circuits. The number of distinct nets per cell is represented by n_c^d, and the number of other cells per distinct net is given by c_d.

Using Eqn. 5.36 and the data of Table 5.7, the values of $\overline{C}$ indicated in the third column of Table 5.8 were obtained. To obtain the actual value of C used in the model for optimized placement, the values of $\overline{C}$ were rounded up to the nearest integer. The rationale is that a portion of a cell connected to a cell c_0 actually takes up a full slot in the clustering of cells around c_0.

Circuit	Cells	Computed $\overline{C}$	Rounded C
i1	33	4.18	5
p1	11	3.34	4
x1	10	6.25	7
i2	23	3.80	4
i3	18	2.76	3
l1	62	13.53	14
d2	20	9.37	10
d1	17	5.82	6
d3	17	8.75	9

Table 5.8 Average number of other cells connected to a cell ($\overline{C}$). The fourth column shows the rounded value of C.

5.6.2 A notion of optimized placement

Consider the checkerboard model, keeping in mind that for any cell c_0, there are, on average, C other cells which are connected to it. Since the placement optimizer seeks to minimize the sum of the bounding boxes of all the nets, it will attempt to place all the C cells in the smallest possible bounding box enclosing the C cells and also the cell c_0. Denote the enclosing bounding box as C_m cells in width and C_s cells in height, where $(C_m C_s) \geq C+1$. A rectangle with minimum perimeter which encloses $C+1$ cells is as close to a square as possible. The notion of minimum perimeter is applied since the objective of the optimizer is to minimize the perimeters of the bounding boxes of the nets. For a given value of C, the dimensions of the enclosing $C_m \times C_s$ rectangle are determined by computing $i = \lceil \sqrt{C+1} \rceil$ (rounding *up* to the nearest integer). If $(i-1)i \geq C+1$, then $C_m = i - 1$ and $C_s = i$.

The placement optimizer encounters a great deal of conflicting demands on the locations of the cells, or *frustration*, when attempting to place all the cells connected to c_0 within the $C_m \times C_s$ rectangle. Consequently, the position of c_0 within the $C_m \times C_s$ rectangle is assumed to be random. Figure 5.12 illustrates three possible rectangles enclosing a cell c_0.

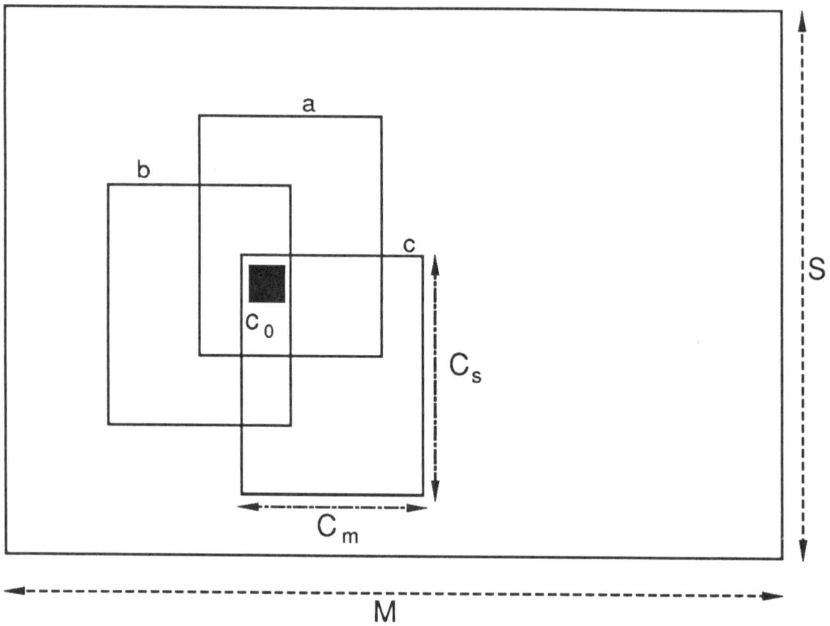

Figure 5.12 Enumerate all possible $C_m \times C_s$ rectangles which enclose the cell c_0. In the figure, three examples are shown: *a*, *b*, and *c*.

The nets connecting to cell c_0 interconnect k–1 other cells located within the $C_m \times C_s$ rectangle. The distribution of these cells is again assumed to be random due to the frustration. The average interconnection length of these nets is then determined by averaging over all possible configurations, or states, of the cells within the $C_m \times C_s$ rectangle.

The strategy is then as follows: (1) for each possible location of c_0, (2) find the positions of all $C_m \times C_s$ rectangles which enclose c_0, and (3) for each such enclosing rectangle, enumerate all cell bounding boxes, with $ms \geq k$, which enclose c_0. Each cell bounding box has an associated interconnection length of $(m + s - 2)$.

Computationally, the enumeration of the states can be made simpler. In Figure 5.13, a particular enclosing rectangle ($C_m = 2$ and $C_s = 3$) is shown.

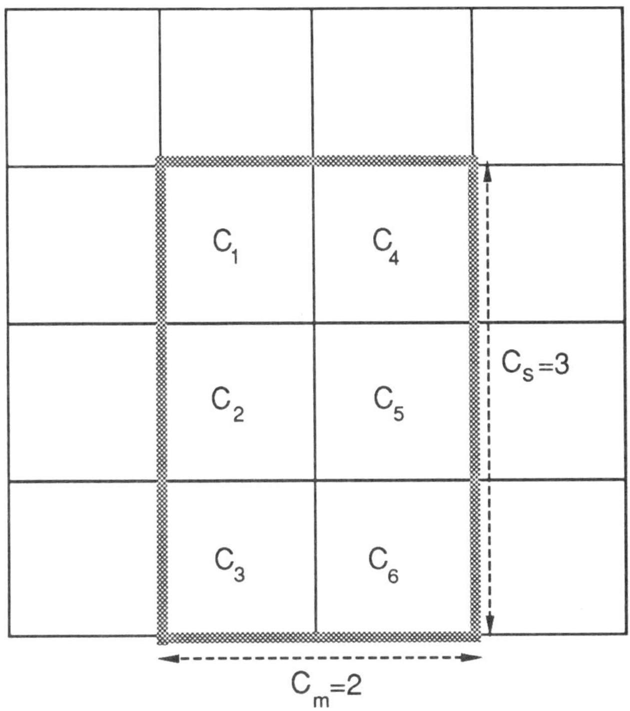

Figure 5.13 Note that the outermost rectangle will only be included as a $C+1$ enclosing rectangle when $c_0 \in \{C_1, C_2, C_3, C_4, C_5, C_6\}$. The internal rectangle represents an $m = 1$ and $s = 2$ cell bounding box. The number of uniquely located $m \times s$ bounding boxes is 4. Given that $M = S = 4$, the number of uniquely located enclosing rectangles is $(4 - 2 + 1) \cdot (4 - 3 + 1) = 6$.

Note that this rectangle will only be included as a $C+1$ enclosing rectangle when $c_0 \in \{C_1, C_2, C_3, C_4, C_5, C_6\}$. Note that each uniquely located $C_m \times C_s$ enclosing rectangle appears equally often ($C_m \cdot C_s$ times). The total number of appearances for a $C_m \times C_s$ enclosing rectangle is therefore given by:

$$(C_m C_s)\,(M - C_m + 1)\,(S - C_s + 1)$$

A parallel argument can be made with respect to the enumeration of the $ms \geq k$ bounding boxes (which enclose c_0) within an enclosing rectangle. The number of appearances is therefore given by:

$$(m\,s)\,(C_m - m + 1)\,(C_s - s + 1)$$

However, as will be seen shortly, the enumeration of $Q_{ms}(k)$ is such that c_0 takes on all possible locations within the $ms \geq k$ cell bounding boxes. This means that the factor (ms) is unnecessary. Hence the term representing the number of appearances for a bounding box of size $ms \geq k$ is reduced to:

$(C_m - m + 1)\,(C_s - s + 1)$

This term also represents the number of uniquely located $m \times s$ bounding boxes.

Furthermore, it is also true that c_0 takes on all possible locations within the enclosing rectangle during the enumeration of $Q_{ms}(k)$. Hence the factor $(C_m C_s)$ is redundant and the term representing the number of enclosing rectangle appearances is therefore reduced to:

$(M - C_m + 1)\,(S - C_s + 1)$

This term also represents the number of uniquely located enclosing rectangles of dimensions $C_m \times C_s$.

The expression for the average interconnection length, $\overline{R}$, for an optimized placement can then be written as:

$$\overline{R} = \frac{\sum_{m=1}^{C_m} \sum_{s=1}^{C_s} (M - C_m + 1)\,(S - C_s + 1)\;Q_{ms}(k)\;(C_m - m + 1)\,(C_s - s + 1)\,(m + s - 2)}{\sum_{m=1}^{C_m} \sum_{s=1}^{C_s} (M - C_m + 1)\,(S - C_s + 1)\;Q_{ms}(k)\;(C_m - m + 1)\,(C_s - s + 1)} \tag{5.37}$$

where $Q_{ms}(k)$ is the number of states which establish the $m \times s$ bounding box for a net of k cells. Seemingly, the term $(M - C_m + 1)\,(S - C_s + 1)$ factors out of Eqn. 5.37; however, as will be indicated later, multiple values of C_m and C_s are considered in some cases.

The computation of $Q_{ms}(k)$ is somewhat different for the optimized placement case. For this situation, the final placement must be feasible. That is, the cells must not overlap one another. Therefore, only one cell may occupy a location in the checkerboard model. While the statistics are thus slightly different, the possible scenarios which establish a given bounding box closely parallels that for the random placement case. Note that there can be no scenario 7 since at least two cells are required to constitute a net.

Scenario 0:

There are no cells in any of the four corners of the $m \times s$ bounding box. In order to *establish* this bounding box, a net must have at least one cell in each of the leftmost and rightmost columns, and at least one cell in each of the bottommost and topmost rows of the bounding box. Note that for this scenario to exist for a given net, the bounding box must be characterized by $m \geq 3$ and $s \geq 3$. Furthermore, the net must consist of $k \geq 4$ cells, and $ms \geq k + 4$ is also required. That is, the bounding box must be large enough to enclose the four empty corners as well as the k cells. It is assumed that there are i_1 cells in the bottom row, i_2 cells in the top row, i_3 cells in the leftmost column, and i_4 cells in the rightmost column of the bounding box. The situation was shown in Figure 5.4.

It follows that there are $(k-i_1-i_2-i_3-i_4)$ cells which do not occupy a bordering position, and therefore occupy the center region of the bounding box. The number of possible arrangements in which there are i_1 cells in the bottom row, i_2 cells in the top row, i_3 cells in the leftmost column, i_4 cells in the rightmost column, and $(k-i_1-i_2-i_3-i_4)$ cells in the center region, is given by:

$$\frac{k!}{i_1!\, i_2!\, i_3!\, i_4!\, (k - i_1 - i_2 - i_3 - i_4)!} \tag{5.38}$$

The extent of the leftmost and rightmost columns is given by (s–2) and the extent of the bottommost and topmost rows is given by (m–2). Since only one cell is permitted to occupy a location in the checkerboard model, there are

$$\frac{(m-2)!}{(m-2-i_1)!}$$

possible arrangements of the i_1 cells in the bottommost row. Because the cells have unique names, it is assumed that the cells are distinguishable. This assumption is consistent with the placement program which will be used to generate comparative experimental results for the average interconnection length.

Similarly, there are

$$\frac{(s-2)!}{(s-2-i_3)!}$$

possible arrangements of the i_3 cells in the leftmost column. Note that there are (ms–$2m$–$2s$+4) slots in the center region, and hence

$$\frac{(ms - 2m - 2s + 4)!}{(ms - 2m - 2s + 4 - (k - i_1 - i_2 - i_3 - i_4))!} \tag{5.39}$$

arrangements of cells in the center region.

Again, a *state* will refer to an arrangement of the cells which establishes the $m \times s$ bounding box. The overall expression for the total number of states, denoted as σ_0, is obtained by summing over all possible values of i_1, i_2, i_3, and i_4. Note that $\sigma_0 = 0$ unless $m \geq 3$, $s \geq 3$, $k \geq 4$, and $ms \geq k + 4$, in which case σ_0 is given by:

$$\sigma_0 = \sum_{i_1=1}^{\min\{k-3,\, m-2\}} \sum_{i_2=1}^{\min\{k-2-i_1,\, m-2\}} \sum_{i_3=1}^{\min\{k-1-i_1-i_2,\, s-2\}} \sum_{i_4=1}^{\min\{k-i_1-i_2-i_3,\, s-2\}} \left\{ \frac{k!}{i_1!\, i_2!\, i_3!\, i_4!} \cdot \right.$$

$$\frac{(m-2)!\,(m-2)!\,(s-2)!\,(s-2)!}{(k-i_1-i_2-i_3-i_4)!\,(m-2-i_1)!\,(m-2-i_2)!\,(s-2-i_3)!\,(s-2-i_4)!} \cdot$$

$$\left. \frac{(ms - 2m - 2s + 4)!}{(ms - 2m - 2s + 4 - (k - i_1 - i_2 - i_3 - i_4))!} \right\} \tag{5.40}$$

Scenario 1:

There are cells in only one of the four corners of the $m \times s$ bounding box. In order to *establish* this bounding box, a net must have at least one cell in one of the corners and at least one cell in the nonadjacent outermost row and column. The situation was depicted in Figure 5.5. Four permutations arise, one generated for each of the different corners which could be the *occupied* corner. Note that for this scenario to exist for a given net, the bounding box must be characterized by $m \geq 3$ and $s \geq 3$. Furthermore, the net must consist of $k \geq 3$ cells, and the relationship $ms \geq k + 3$ must be obeyed. With respect to Figure 5.5, it is assumed that there are i_1 cells in the bottom row, i_2 cells in the leftmost column, and i_3 cells in the upper right corner.

The expression for the total number of states is denoted by σ_1. This expression was developed using arguments which closely parallel those presented for the case of scenario 0. Note that $\sigma_1 = 0$ unless $m \geq 3$, $s \geq 3$, $k \geq 3$, and $ms \geq k + 3$, in which case σ_1 is given by:

$$\sigma_1 = \sum_{i_1=1}^{\min\{k-2,\, m-2\}} \sum_{i_2=1}^{\min\{k-1-i_1,\, s-2\}} \frac{4k!\ (m-2)!\ (s-2)!\ (ms-m-s)!}{i_1!\ i_2!\ (k-i_1-i_2-1)!\ (m-2-i_1)!\ (s-2-i_2)!\ (ms-m-s-(k-i_1-i_2-1))!} \tag{5.41}$$

Scenario 2A:

There are cells in exactly two opposite corners of the $m \times s$ bounding box. In order to *establish* this bounding box, a net must have at least one cell in each of the corners. The situation was depicted in Figure 5.6. Two permutations arise, one generated for each of the two sets of opposite-corner pairs which are occupied. Note that for this scenario to exist for a given net, the bounding box must be characterized by $m \geq 2$ and $s \geq 2$. Furthermore, the net must consist of $k \geq 2$ cells, and the expression $ms \geq k + 2$ must be obeyed. It is assumed that there are i_1 cells in one of the corners and i_2 cells in the other corner. The expression for the total number of states is denoted by σ_{2A}. Note that $\sigma_{2A} = 0$ unless $m \geq 2$, $s \geq 2$, $k \geq 2$, and $ms \geq k + 2$, in which case σ_{2A} is given by:

$$\sigma_{2A} = \frac{2k!\ (ms-4)!}{(k-2)!\ (ms-4-(k-2))!} \tag{5.42}$$

Scenarios 2B and 2C:

There are cells in exactly two adjacent corners of the $m \times s$ bounding box. In order to *establish* this bounding box, a net must have at least one cell in each of the two corners. The situation was depicted in Figure 5.7. Scenario 2B denotes the situation when the two occupied corners lie on the same outer row, as depicted in Figure 5.7. For this case, two permutations arise, one generated in which the occupied corners are in the topmost row and one in which they are in the bottommost row. Note that for this scenario to exist for a given net, the bounding box must be characterized by $m \geq 3$ and $s \geq 2$. Furthermore, the net must consist of $k \geq 3$ cells, and $ms \geq k + 2$ is required. With respect to Figure 5.7, it is assumed that there are i_1 cells in one of the outermost rows, i_2 cells in one of the corners and i_3 cells in the other occupied corner. The expression for the total number of states is denoted by σ_{2B}. Note that $\sigma_{2B} = 0$ unless $m \geq 3$, $s \geq 2$, $k \geq 3$, and $ms \geq k + 2$, in which case σ_{2B} is given by:

$$\sigma_{2B} = \sum_{i_1 = 1}^{\min\{k-2,\, s-2\}} \frac{2k!\,(m-2)!\,(ms-m-2)!}{i_1!\,(k-i_1-2)!\,(m-2-i_1)!\,(ms-m-2-(k-i_1-2))!} \tag{5.43}$$

Scenario 2C:

Scenario 2C denotes the situation when the two occupied corners lie on the same outer column. For this scenario to exist for a given net, the bounding box must be characterized by $m \geq 2$ and $s \geq 3$. Furthermore, the net must consist of $k \geq 3$ cells. That is, $\sigma_{2C} = 0$ unless $m \geq 2$, $s \geq 3$, $k \geq 3$, and $ms \geq k + 2$, in which case σ_{2C} is given by:

$$\sigma_{2C} = \sum_{i_1 = 1}^{\min\{k-2,\, s-2\}} \frac{2k!\,(s-2)!\,(ms-s-2)!}{i_1!\,(k-i_1-2)!\,(s-2-i_1)!\,(ms-s-2-(k-i_1-2))!} \tag{5.44}$$

Scenario 3:

For a net to establish this bounding box, there must be at least one cell in exactly three of the corners. The situation was depicted in Figure 5.8. There are i_1, i_2, and i_3 cells in each of the three corners, respectively. For this scenario to exist for a given net, the bounding box must be characterized by $m \geq 2$ and $s \geq 2$. Furthermore, the net must consist of $k \geq 3$ cells. That is, $\sigma_3 = 0$ unless $m \geq 2$, $s \geq 2$, $k \geq 3$, and $ms \geq k + 1$, in which case σ_3 is given by:

$$\sigma_3 = \frac{4k!\,(ms-4)!}{(k-3)!\,(ms-4-(k-3))!} \tag{5.45}$$

Scenario 4:

For a net to establish this bounding box, there must be at least one cell in each of the four of the corners. The situation was depicted in Figure 5.9. There are i_1, i_2, i_3, and i_4 cells in each of the four corners, respectively. $\sigma_4 = 0$ unless $m \geq 2$, $s \geq 2$, $k \geq 4$, and $ms \geq k$, in which case σ_4 is given by:

$$\sigma_4 = \frac{k!\,(ms-4)!}{(k-4)!\,(ms-4-(k-4))!} \tag{5.46}$$

Scenario 5:

Scenarios 5 and 6 represent the bounding boxes which are actually not rectangles, but rather rows or columns, respectively, in the checkerboard model. For a net to establish a scenario 5 bounding box, the cells belonging to the net must all lie in one row of the checkerboard model. There must be at least one cell at each endpoint of the $m \times 1$ bounding box, or row. The situation was illustrated in Figure 5.10. There are i_1 cells at one end and i_2 cells at the other end. $\sigma_5 = 0$ unless $s = 1$ and $m \geq k$, in which case σ_5 is given by:

$$\sigma_5 = \frac{k!\,(m-2)!}{(k-2)!\,(m-2-(k-2))!} \tag{5.47}$$

Scenario 6:

For a net to establish a scenario 6 bounding box, the cells belonging to the net must all lie in one column of the checkerboard model. There must be $i_1 \geq 1$ cells at one endpoint and $i_2 \geq 1$ cells at the other endpoint of the $1 \times s$ bounding box, or column. $\sigma_6 = 0$ unless $m = 1$ and $s \geq k$, in which case σ_6 is given by:

$$\sigma_6 = \frac{k!\ (s-2)!}{(k-2)!\ (s-2-(k-2))!} \tag{5.48}$$

The expressions for the total number of states $Q_{ms}(k)$ for given values of k, m, and s are summarized in Table 5.9. Note that for a given set of values of k, m, and s, only one row in the table is relevant.

Case	k	m	s	The Total Number of States $Q_{ms}(k)$:
1	$4 \leq k \leq ms-4$	≥ 3	≥ 3	$\sigma_0 + \sigma_1 + \sigma_{2A} + \sigma_{2B} + \sigma_{2C} + \sigma_3 + \sigma_4$
2	$4 \leq k = ms-3$	≥ 3	≥ 3	$\sigma_1 + \sigma_{2A} + \sigma_{2B} + \sigma_{2C} + \sigma_3 + \sigma_4$
3	$4 \leq k = ms-2$	≥ 3	≥ 3	$\sigma_{2A} + \sigma_{2B} + \sigma_{2C} + \sigma_3 + \sigma_4$
4	$4 \leq k = ms-1$	≥ 3	≥ 3	$\sigma_3 + \sigma_4$
5	$4 \leq k = ms$	≥ 3	≥ 3	σ_4
6	$4 \leq k \leq ms-2$	≥ 3	2	$\sigma_{2A} + \sigma_{2B} + \sigma_3 + \sigma_4$
7	$4 \leq k = ms-1$	≥ 3	2	$\sigma_3 + \sigma_4$
8	$4 \leq k = ms$	≥ 3	2	σ_4
9	$4 \leq k \leq ms-2$	2	≥ 3	$\sigma_{2A} + \sigma_{2C} + \sigma_3 + \sigma_4$
10	$4 \leq k = ms-1$	2	≥ 3	$\sigma_3 + \sigma_4$
11	$4 \leq k = ms$	2	≥ 3	σ_4
12	$k = 4$	2	2	σ_4
13	$2 \leq k \leq \text{m}$	>1	1	σ_5
14	$2 \leq k \leq s$	1	>1	σ_6
15	3	≥ 3	≥ 3	$\sigma_1 + \sigma_{2A} + \sigma_{2B} + \sigma_{2C} + \sigma_3$
16	3	≥ 3	2	$\sigma_{2A} + \sigma_{2B} + \sigma_3$
17	3	2	≥ 3	$\sigma_{2A} + \sigma_{2C} + \sigma_3$
18	3	2	2	σ_3
19	2	≥ 2	≥ 2	σ_{2A}

Table 5.9 The optimized placement problem. There are 19 cases for the computation of $Q_{ms}(k)$ for nets of k cells to establish a bounding box of size $m \times s$.

The total number of states $Q_{ms}(k)$ for a net consisting of k cells is readily computed based on the appropriate row of Table 5.9. The expression for the average interconnection length, $\overline{R}$, for an optimized placement was given by Eqn. 5.37. This result is augmented here to include a range of $C_m \times C_s$ enclosing rectangle sizes, as well as an accommodation for nets of various numbers of cells.

$$\overline{R} = \frac{\sum_{C_m} \sum_{C_s} \sum_{k=2}^{k_{max}} \sum_{m=1}^{C_m} \sum_{s=1}^{C_s} N(k)\,(M-C_m+1)(S-C_s+1)\,Q_{ms}(k)\,(C_m-m+1)(C_s-s+1)(m+s-2)}{\sum_{C_m} \sum_{C_s} \sum_{k=2}^{k_{max}} \sum_{m=1}^{C_m} \sum_{s=1}^{C_s} N(k)\,(M-C_m+1)(S-C_s+1)\,Q_{ms}(k)\,(C_m-m+1)(C_s-s+1)} \tag{5.49}$$

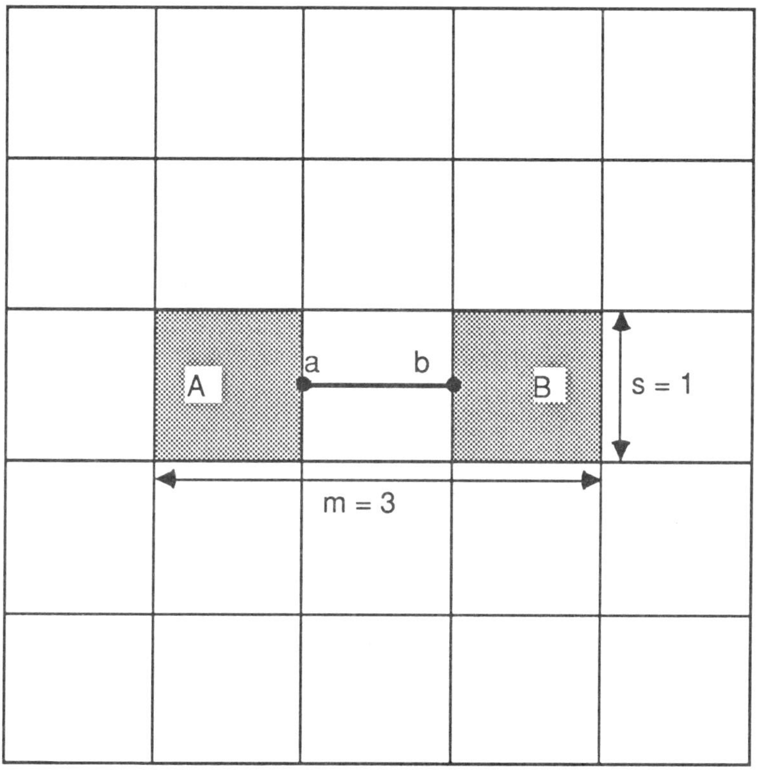

Figure 5.14 A two-pin net interconnecting two cells A and B. The value of $(m + s - 2)$ is 2. However, a placement optimizer is usually allowed to adjust the orientation of the cells in order to minimize the interconnection length. The optimizer would therefore seek to adjust the cell orientations such that the pins belonging to each net are as close as possible. It would typically be able to achieve a length of 1 for this example.

If the pin positions are not located at the center of the cells, but rather at the edges, Eqn. 5.49 will not be able to accurately reflect the performance of a placement optimizer.

For example, consider a two-pin net interconnecting two cells, A and B, as illustrated in Figure 5.14. The value of $(m + s - 2)$ is 2. However, a placement optimizer is usually allowed to adjust the orientation of the cells in order to minimize the interconnection length. The optimizer would therefore seek to adjust the cell orientations such that the pins belonging to each net are as close as possible. It would typically be able to achieve a length of 1 for the example of Figure 5.14. This effect is modeled by replacing the factor $(m + s - 2)$ by the following in Eqn. 5.49:

$$\{ \max \{ 0, m - 2 \} + \max \{ 0, s - 2 \} \} \tag{5.50}$$

The TimberWolfMC placement optimizer presented in the previous chapter appends borders around the contours of the cells so as to maintain sufficient interconnect space between the cells. In this case, Eqn. 5.50 is not the appropriate expression since pins on adjacent cells must be separated by at least a distance β, as illustrated in Figure 5.15.

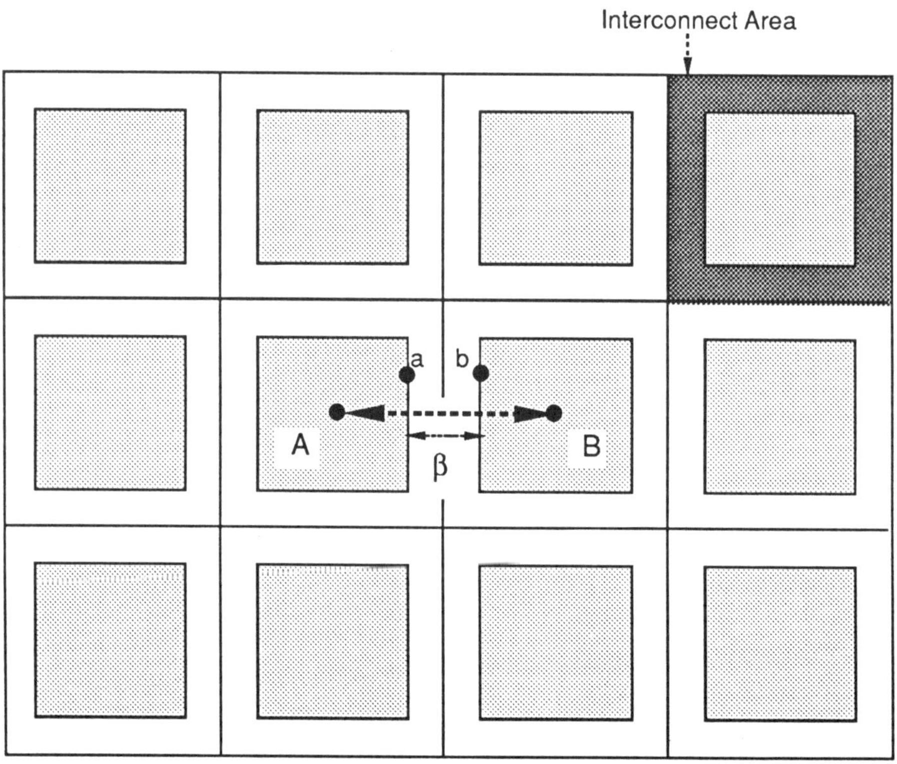

Figure 5.15 The TimberWolfMC placement optimizer presented in the previous chapter appends borders around the contours of the cells so as to maintain sufficient interconnect space between the cells. In this case, Eqn. 5.50 is not the appropriate expression since pins on adjacent cells must be separated by at least a distance β as indicated.

A more accurate expression is the following, where L represents the interconnection length for a net:

$$L(m, s, \beta) = \{ \max \{ 0, m - 2 + \beta \} + \max \{ 0, s - 2 + \beta \} \} \quad (5.51)$$

With the use of Eqn. 5.51, the complete expression for $\overline{R}$ becomes:

$$\overline{R} = \frac{\sum_{C_m} \sum_{C_s} \sum_{k=2}^{k_{max}} \sum_{m=1}^{C_m} \sum_{s=1}^{C_s} N(k)\,(M-C_m+1)(S-C_s+1)\,Q_{ms}(k)\,(C_m-m+1)(C_s-s+1)\,L(m,s,\beta)}{\sum_{C_m} \sum_{C_s} \sum_{k=2}^{k_{max}} \sum_{m=1}^{C_m} \sum_{s=1}^{C_s} N(k)\,(M-C_m+1)(S-C_s+1)\,Q_{ms}(k)\,(C_m-m+1)(C_s-s+1)} \quad (5.52)$$

5.6.3 The enclosing $C_m \times C_s$ rectangles

Recall that for any cell c_0, there are, on average, C other cells which are connected to it. The placement optimizer seeks to place all the C cells in the smallest possible bounding box enclosing the C cells and also the cell c_0. A rectangle with minimum perimeter and with minimum area which encloses the C+1 cells was determined by computing

$$i = \lceil \sqrt{C+1} \rceil$$

However, if $(i-1)\, i \geq C+1$, then $C_m = i - 1$ and $C_s = i$.

The consideration of multiple sizes of the enclosing rectangles was done to acknowledge the degree of frustration present in the placement problem. However, it was assumed that the placement optimizer will usually be able to place the C+1 cells within the enclosing $C_m \times C_s$ rectangle. A second enclosing rectangle size was used when $(C_m C_s) \approx (C + 1)$. That is, it was assumed that the placement optimizer would, at least occasionally, have some difficulty enclosing all of the C+1 cells in a rectangle which just barely accommodates them all. In particular, for $(C_m C_s) \leq ((C + 1) + 1)$, a second enclosing rectangle was also considered. The size of this second rectangle was taken to the next largest rectangle which was nearly a square. That is, if $C_m < C_s$, then C_m was incremented by one to determine the second rectangle. On the other hand, if $C_m = C_s$, then C_s was incremented by one to determine the second rectangle. Table 5.10 shows the sets of enclosing rectangles which were applied for the nine industrial circuits.

Circuit	Cells	C	Associated $C_m \times C_s$ Enclosing Rectangles
i1	33	5	2×3 , 3×3
p1	11	4	2×3 , 3×3
x1	10	7	3×3 , 3×4
i2	23	4	2×3 , 3×3
i3	18	3	2×2 , 2×3
l1	62	14	4×4 , 4×5
d2	20	10	3×4 , 4×4
d1	17	6	3×3
d3	17	9	3×4

Table 5.10 Average number of other cells connected to a cell (C), and the dimensions of the enclosing rectangles associated with the value of C.

5.7 Results

Given a circuit, Eqns. 5.32 and 5.52 enable the computation of the predicted value for ρ, the average interconnection length for a random placement divided by the average interconnection length for an optimized placement.

$$\rho = \frac{\overline{R}_r}{\overline{R}} \tag{5.53}$$

The average total interconnect length L_R for random placements can be measured easily by setting $T = \infty$ in the placement algorithm of the previous chapter. Then the final total interconnect length can be estimated by

$$L_o = \frac{L_R}{\rho} \tag{5.54}$$

To test the model for optimized placement, the experimental values of ρ were compared to the predicted values derived from Eqns. 5.32 and 5.52. The TimberWolfMC placement optimization algorithm of the previous chapter was used to do the measurements. The results for the nine industrial macro/custom cell circuits are shown in Table 5.11.

Circuit	Cells	β	Predicted ρ	Measured ρ	Deviation (%)
i1	33	0.2	5.0	5.3	6
p1	11	0.2	2.0	2.1	5
x1	10	0.2	1.95	1.93	1
i2	23	0.4	2.9	2.6	10
i3	18	0.4	2.9	2.6	10
l1	62	0.15	2.2	2.37	7
d2	20	0.1	2.2	2.5	13
d1	17	0.6	1.7	1.8	5
d3	17	0.06	1.8	1.7	5
avg.					7

Table 5.11 Comparison of predicted and measured total interconnect length reduction ratios (ρ). β indicates the percentage of the center-to-center cell spacing due to interconnect area estimation.

Note that the average interconnection length reduction ratios for the predicted method were typically within seven percent of the measured ratios. The results of Table 5.11 imply that the checkerboard model is sufficient to derive accurate interconnection models for both random and optimized placement, even when the cells are of widely varying size and shape. In fact, Eqn. 5.32 gives the *precise* value of $\overline{R}_r$ for circuits with multi-pin nets which obey the checkerboard model. *Island-style* gate arrays, presented in the Appendix, obey this model. Hence, for these circuits, the experimental measurements of the average interconnection length for random placements *must* coincide with the so-called predicted value given by Eqn. 5.32. If not, the error lies with the measurements and not with the theoretical model.

It is thus now possible to calculate the ratio ρ of the average interconnection length for random placement over optimized placement. The dynamic interconnect-area estimator, described in the next chapter, depends heavily on an accurate prediction of the final total interconnect length for an optimized placement. While it would be possible to obtain such an estimate directly from Eqn. 5.52, the fact that the circuits often have cells of widely varying sizes and shapes, including nonrectangular shapes, the following procedure is used instead to *tune* the model to the nuances of a circuit. First, measure the random total interconnect length L_R for the circuit. Then compute the final total interconnect length by using Eqn. 5.54. This procedure effectively forces any modeling inaccuracies (due to the nonuniform cells sizes and shapes) to cancel out during the computation of ρ, since the inaccuracies will appear for both the random (numerator) and the optimized (denominator) placement cases.

Chapter 6

Interconnect-Area Estimation for Macro Cell Placements

6.1 Introduction

In Chapter 3 it was shown that simulated annealing was an effective algorithm upon which to base a standard cell placement and global routing package. Furthermore, in Chapter 4 it was shown that simulated annealing was also effective for the generalized macro/custom cell placement problem. For this latter layout style, the cells are permitted to have any rectilinear shape. Furthermore, the cells may have fixed geometry including pin locations (macro cells) or the cells may have an estimated area with a specified aspect-ratio range, and with pins that need to be placed (custom cells).

For macro and custom cells, it is usually the case that there are pins on all of the edges of the cells. It is therefore necessary to allocate interconnect space around each cell. Failure to allocate the correct amount invariably results in substantial placement alteration following global routing and/or during detailed routing. These placement modifications usually result in substantial total interconnect length increases as well as chip area increases because of the introduction of sizable empty areas on the chip. Furthermore, attempting to find the most suitable cell–placement manipulation is indeed a challenging problem. Solving this type of problem requires either a placement–refinement algorithm or a spacer.[1] Since the placement adjustments are local in nature, suboptimal solutions are likely.

Unfortunately, it is very difficult to compute the precise amount of interconnect space needed along each edge of a cell without performing a global routing step. Furthermore, it is even more difficult to compute the interconnect space during an intermediate

[1] J. Burns and R. Newton, "SPARCS: A New Constraint-Based IC Symbolic Layout Spacer," (1986).

point of a simulated annealing approach in which a penalty function is used to limit the amount of cell overlapping. During such an intermediate point, it is likely that several cases of cell overlapping are present. This makes a channel definition procedure very difficult to perform. Thus, while it would be best to maintain a global routing of the placement at any given point in the simulated annealing process, such a procedure is very difficult at best, and surely very CPU–time intensive.

It is possible to define three general categories of interconnect–area estimation problems: (1) The specifics of the cell placement are not of concern. For each circuit examined, it is assumed that the *quality* of the placement will be comparable. For this category, the interconnect model does not incorporate the detailed placement information. The method by Heller, Mikhail, and Donath has been very successful for solving problems in this category.[1] (2) It is desired to compute the expected interconnect area for a single, specific cell placement. Since the computation will be performed only one time for a given circuit, the computation time is less crucial than in the next category. The model developed by El Gamal and Syed has been successful in solving problems in this category.[2] (3) The interconnect area must be estimated for very many placements of the same circuit (perhaps one million or more). It is required that the model take into account the specifics of the cell placement, and also that the computation time needed to compute the estimate must be relatively short. This category will be referred to as the *dynamic* interconnect–area estimation problem since the placement of the cells is, obviously, not static.

The method introduced by Heller, Mikhail, and Donath for solving problems in category 1 appears not to be suitable for the dynamic estimation problem since the specifics of the cell placement are not taken into account. Also, the method introduced by El Gamal and Syed for solving problems in category 2 is apparently not suitable for the dynamic estimation problem since it appears that the computational requirements are too severe to handle one million or more placements. Their method also assumes that the pin locations on the various cells have a statistical distribution. However, in the problems of interest here, the pin locations for a given cell are either precisely known (macro cells) or approximately known (custom cells).

The approach introduced here estimates the interconnect area needed for each edge of a cell. The estimate is based on three factors: (1) the *average net traffic*, which is an estimate of the average number of interconnections passing through a cross section of any channel, (2) the relative position of the edge on the chip (routing channels in the center of the chip tend to be wider than those channels near the boundary of the chip), and (3) the relative pin density of the edge (that is, in comparison to the average pins per unit edge length). Note that factor (2) above must be updated each time a cell is moved during the simulated annealing process. The CPU time required to update this factor is minimal. Factors (1) and (3) can be determined at the outset and stored.

---

[1] W. Heller, W. Mikhail, and W. Donath, "Prediction of Wiring Space Requirements for LSI," (1978).

[2] A. El Gamal and Z. Syed, "A Stochastic Model for Interconnections in Custom Integrated Circuits," (1981).

The dynamic interconnect–area estimator is a key element of TimberWolfMC, the macro/custom cell placement and global routing package described in Chapter 4. The pin placement specification permitted by TimberWolfMC is fairly extensive and must be accommodated by the dynamic estimator. Pins on macro cells must be given specific locations, either inside, on, or outside the cell boundaries (edges). However, pins on custom cells may be specified in several possible ways: (1) a pin may be given a particular fixed location, (2) a pin may be assigned to a particular edge or edges of a cell, (3) a pin may belong to a group of pins which may be assigned to a particular edge or edges of a cell, or (4) a pin may belong to a group of pins which is assigned a particular sequence ordering as well as a particular edge or edges of a cell. The placement of a single pin, a group of pins, or a sequence of pins may be specified as being confined to one cell edge, two cell edges, or any of the edges.

In the simulated annealing approach used in TimberWolfMC, sufficient interconnect space between the cells is maintained by appending a border around the contours of each cell. The area occupied by each rectilinear cell is initially represented as a set of one or more nonoverlapping rectangular tiles. Subsequently, the edges of each tile are expanded outward in accordance with the expected interconnect-area requirements. That is, the left edge of each tile will be extended to the left by an amount related to: (1) its horizontal position with respect to the center of the core, (2) its top and bottom endpoint positions, and (3) its relative pin density. Similarly, the right edge of each tile will be extended to the right by an amount in accordance with the three factors outlined for the left edges. Also, the top edges of the tiles are expanded upward in accordance with (1) its vertical position, (2) its left and right endpoint positions, and (3) its relative pin density. Similarly, the bottom edge of each tile will be extended downward by an amount in accordance with the three factors outlined for the top edges.

The respective outward expansions of the tile edges are updated each time a cell participates in the generation of a new state in the simulated annealing approach. Hence the estimation of the interconnect area attributable to a given edge of a cell is a dynamic quantity, depending on the location of the edge at any given time. If a cell is moved from a corner of the core area towards the center of the core area, then the effective area of the cell will increase. Similarly, if a cell is moved from the center towards a corner of the core area, then the effective area of the cell will decrease.

The effectiveness of the estimator is demonstrated by comparing the final chip area as achieved by the simulated annealing algorithm using the estimator versus the final chip area following a global routing step. Also, the total interconnect length was compared for the two cases, since a disparity would indicate substantial cell movement was necessary to accommodate the global routing. Test results are reported for nine industrial macro/custom cell circuits. The final chip areas and the final total interconnect lengths are within a few percent of each other for all nine examples. This dynamic estimator has enabled the generation of placements which require very little placement modification during or after detailed routing.

Six sections comprise the remainder of this chapter. The next four sections derive the three major components of the dynamic interconnect–area estimator. Section 6.2 presents the derivation of the *average net traffic*, which is used to compute the baseline (or average) channel widths. Section 6.3 presents the derivation of the channel–width modulation based on the relative position of the channel. In Section 6.4, the results of Section 6.3 are used in associating the estimated interconnect area with the cell edges instead of the channels. In Section 6.5, interconnect–area modulation as a function of *relative pin density* is derived. Section 6.6 describes the implementation of the dynamic interconnect–area estimator. This chapter concludes with a presentation of the results in Section 6.7.

6.2 Interconnect–Area Estimation Based on Average Net Traffic

The average net traffic, or N_T, is an estimate of the average number of interconnections passing through a cross-section of a channel. The value for N_T is computed by taking an estimate of the final total interconnect length and dividing by an estimate of the sum of the lengths of the channels.

The results of the previous chapter are used to compute the value of L_O, or final total interconnect length. First, the random total interconnect length (L_R) for a circuit is determined. This can be ascertained by setting $T = \infty$ (implying a 100 percent acceptance rate) in the placement algorithm described in Chapter 4, and then visiting a statistically significant number of new states while measuring the average total interconnect length.

Second, the final total interconnect length can be accurately estimated by using the results of the previous chapter:

$$L_o = \frac{L_R}{\rho} \tag{6.1}$$

where ρ is the predicted ratio of the average interconnection length for random placement ($\overline{R}_r$) over optimized placement ($\overline{R}$) for the particular circuit. That is,

$$\rho = \frac{\overline{R}_r}{\overline{R}} \tag{6.2}$$

Thus the value of L_O is readily computed from the measured total interconnect length for random placements and the expressions derived in the previous chapter.

Next, consider a rectangular region which is subdivided into a set of nonoverlapping rectangular tiles (or cells) which completely cover the rectangular region. An example is shown in Figure 6.1, where eleven cells fully cover a rectangular region.

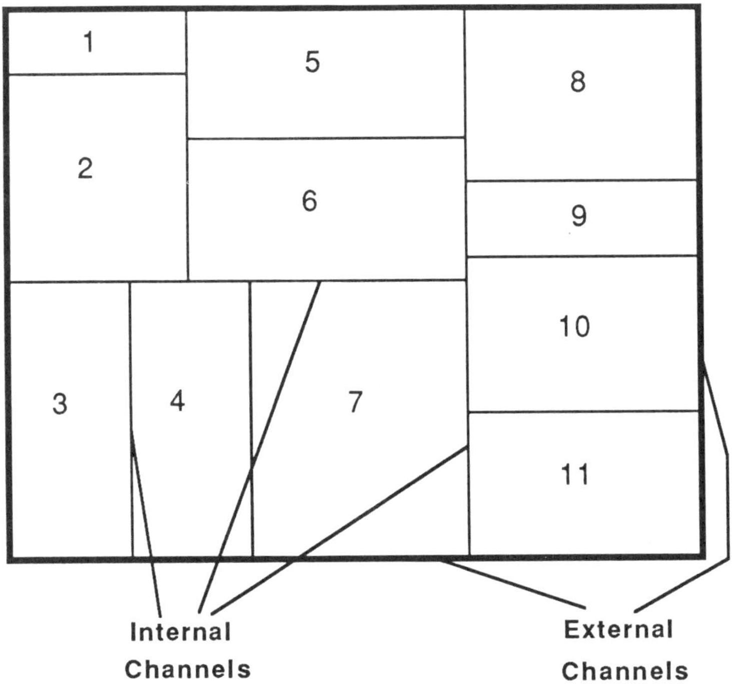

Figure 6.1 An internal channel refers to a channel which lies between two cells. An external channel does not lie between two cells, but rather lies on one of the four edges of the rectangular region.

In this model, the cells fully abut one another and the channels have zero thickness. The term *internal* channel will refer to a channel which lies between two cells. On the other hand, an *external* channel does not lie between two cells, but rather lies on one of the four edges of the rectangular region. Examples of internal and external channels are indicated in Figure 6.1.

The total channel length for a cell placement will be approximated by the sum of the lengths of the internal channels. It is almost always the case that the shortest possible routes are used for all nets. Hence, it would not be expected that the external channels would be utilized to a great extent for routing. The sum of the lengths of the internal channels will be referred to as C_L. For the model of Figure 6.1, note that summing the perimeters of all of the cells counts each internal channel edge twice; however, the external channel edges are only counted once. Let N_c represent the number of cells and define $p(c)$ as a function returning the perimeter of cell c, where $c \in \{1, 2, \cdots, N_c\}$. Further, let P_c represent the perimeter of the rectangular region (equivalently, the core or bounding box) containing the cells. Then the expression for the total internal channel length is:

$$C_L = \frac{1}{2}\left[\sum_{c=1}^{N_c} p(c)\right] - \frac{1}{2}P_c \tag{6.3}$$

The use of Eqns. 6.1 and 6.3 then yield the estimate for the average net traffic:

$$N_T = \frac{L_o}{C_L} \tag{6.4}$$

Defining t_s as the center–to–center spacing required between wiring tracks in the channels, the average channel width (C_W) is given by

$$C_W = N_T\, t_s \tag{6.5}$$

The value of C_W will be used as the baseline expectation of the width of any channel, modulated by three functions which are described in the next two sections.

6.3 Baseline Channel Width Modulation Based on Channel Position

In macro/custom cell layout, it is often the case that the shortest possible route is used for a net. Consequently, it would be expected that the thicknesses of the channels would be greater nearer the center of the core area. This effect is modeled by multiplying C_W by a modulation function, whose value reflects the position of the channel. This modulation function is implemented as the composite of two modulation functions, one reflecting the vertical position of a channel edge and the other reflecting the horizontal position.

A two–piece linear function is used for each of the two modulation functions, in which the maximum value of the function occurs for position zero and the minimum values occur at each edge of the core region. For the horizontal modulation function, $f_x()$, the maximum value of the function is represented by M_x and the minimum value of the function is represented by B_x. Similarly, for the vertical modulation function, $f_y()$, the maximum value of the function is represented by M_y and the minimum value of the function is represented by B_y. The nature of the two modulation functions is shown graphically in Figure 6.2. The center of the core region is set at $x = 0$ and $y = 0$, with the width of the core represented by W and the height of the core represented by H.

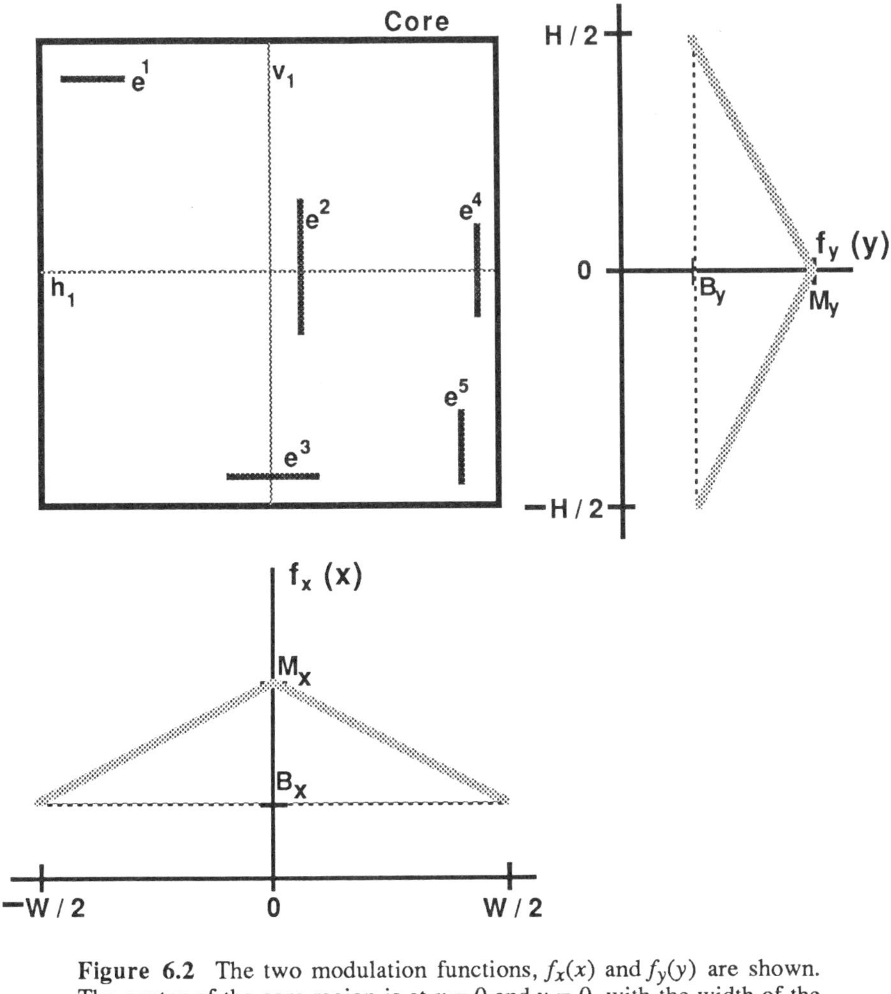

Figure 6.2 The two modulation functions, $f_x(x)$ and $f_y(y)$ are shown. The center of the core region is at $x = 0$ and $y = 0$, with the width of the core equal to W and the height equal to H. Five channel edges are shown. Edge e^1 is characterized by $f_x(e^1) \approx B_x$ and $f_y(e^1) \approx B_y$. Edge e^2 is characterized by $x_i = y_i = 0$. Hence, $f_x(e^2)\, f_y(e^2) = M_x\, M_y$. Edge e^3 has $x_i = 0$ and $y_i \approx -0.5H$. Hence, $f_x(e^3)\, f_y(e^3) \approx M_x\, B_y$. The modulation factors for edge e^4 are very similar to those of edge e^3 in the respect that one factor is assigned maximum weight and the other is assigned near minimum weight. For this edge, $f_x(e^4)\, f_y(e^4) \approx B_x\, M_y$. Edge e^5 is such that $x_i \approx 0.5W$ and $y_i \approx -0.5H$. The modulation factors for this edge are therefore assigned minimum weights: $f_x(e^5)\, f_y(e^5) \approx B_x\, B_y$.

The expression for the horizontal modulation function is given below:

$$f_x(x) = M_x - \left[\frac{M_x - B_x}{0.5W}\right] |x| \quad (6.6)$$

where $-0.5W \le x \le 0.5W$. Note that $f_x()$ has a maximum value of M_x for $x = 0$ and a minimum value of B_x for $x = \pm 0.5W$.

The expression for the vertical modulation function is given below:

$$f_y(y) = M_y - \left[\frac{M_y - B_y}{0.5H}\right] |y| \quad (6.7)$$

where $-0.5H \le y \le 0.5H$. Again note that $f_y()$ has a maximum value of M_y for $y = 0$ and a minimum value of B_y for $y = \pm 0.5H$.

The functional forms for $f_x()$ and $f_y()$ were chosen based on the observation of manual layouts of industrial macro cell chips. For two layers of interconnect, the channels near the center of the core usually were approximately twice as wide as channels located near the middle of the sides of the core, and approximately four times as wide as channels located near the corners of the core. Consequently, typical selections are: $M_x = M_y \approx 2$ and $B_x = B_y \approx 1$.

Each channel edge, whether a vertical channel edge or a horizontal channel edge, is multiplied by both modulation functions in determining its expected width. The expression for the expected channel width for channel edge i is represented by e_w^i, and is shown below:

$$e_w^i = C_W \; f_x(x_i) \; f_y(y_i) \quad (6.8)$$

where x_i and y_i are the *effective* x and y locations of channel edge i.

The values of x_i and y_i for channel i are determined as follows: Each channel edge, as in Figure 6.2, is described by two pairs of coordinates. One pair, (x_1^i, y_1^i), is the location of one end of channel edge i and the other pair, (x_2^i, y_2^i), is the location of the other end.

There are then two possibilities. First, suppose that edge i is a vertical channel edge. In this case, $x_i = x_1^i = x_2^i$. The determination of y_i is a bit more complicated. If vertical edge i crosses the horizontal center of the core, represented by line h_1 in Figure 6.2, then $y_i = 0$ is assigned. This indicates that maximum weight in the vertical modulation function is assigned since this channel crosses the horizontal center of core. On the other hand, if vertical edge i does not cross line h_1, y_i is computed as follows (noting that the center of the core is at (0, 0)):

$$y_i = \min \{ \, | \, y_1^i \, | \, , \, | \, y_2^i \, | \, \} \quad (6.9)$$

That is, the endpoint of the vertical edge which is nearest line h_1 dictates the value of y_i. The maximum in congestion of a channel may occur at any point along the channel edge. Since the point nearest the center of the core is the most likely to encompass the maximum congestion, the computation of y_i reflects this concept.

Second, suppose that edge i is a horizontal channel edge. In this case, $y_i = y_1^i = y_2^i$. The determination of x_i is more complicated for this case. If horizontal edge i crosses the vertical center of the core, represented by line v_1 in Figure 6.2, then $x_i = 0$ is assigned. This indicates that maximum weight in the horizontal modulation function is assigned since this channel crosses the vertical center of core. On the other hand, if horizontal edge i does not cross line v_1, x_i is computed as follows:

$$x_i = \min\{\,|x_1^i|\,,\,|x_2^i|\,\} \tag{6.10}$$

That is, the endpoint of the horizontal edge which is nearest line v_1 dictates the value of x_i. Again, Eqn. 6.10 reflects the fact that the maximum in congestion in a channel is expected to be at the end of the channel nearest the center of the core.

Based on the above analysis, including the development of Eqns. 6.9 and 6.10, the expressions for x_i and y_i are given below:

$$x_i = \begin{cases} 0 & \text{if } x_1^i \geq 0 \text{ and } x_2^i \leq 0 \\ 0 & \text{if } x_1^i \leq 0 \text{ and } x_2^i \geq 0 \\ \min\{\,|x_1^i|\,,\,|x_2^i|\,\} & \text{if } x_1^i > 0 \text{ and } x_2^i > 0 \\ \min\{\,|x_1^i|\,,\,|x_2^i|\,\} & \text{if } x_1^i < 0 \text{ and } x_2^i < 0 \end{cases} \tag{6.11}$$

$$y_i = \begin{cases} 0 & \text{if } y_1^i \geq 0 \text{ and } y_2^i \leq 0 \\ 0 & \text{if } y_1^i \leq 0 \text{ and } y_2^i \geq 0 \\ \min\{\,|y_1^i|\,,\,|y_2^i|\,\} & \text{if } y_1^i > 0 \text{ and } y_2^i > 0 \\ \min\{\,|y_1^i|\,,\,|y_2^i|\,\} & \text{if } y_1^i < 0 \text{ and } y_2^i < 0 \end{cases} \tag{6.12}$$

In Figure 6.2, five channel edges are shown. Edge e^1 is characterized by $f_x(e^1) \approx B_x$ and $f_y(e^1) \approx B_y$. That is, channel edge e^1 is expected to have minimum width since it is nearly as far as possible from the vertical and horizontal centers of the core. On the other hand, edge e^2 is characterized by the following effective x and y locations: $x_i = y_i = 0$. Hence, $f_x(e^2)\, f_y(e^2) = M_x\, M_y$.

Edge e^3 has effective x and y locations which are $x_i = 0$ (since the horizontal span of the channel edge crosses the vertical center of the core) and $y_i \approx -0.5H$. Hence, $f_x(e^3)\, f_y(e^3) \approx M_x\, B_y$. The modulation factors for edge e^4 are very similar to those of edge e^3 in the respect that one factor is assigned maximum weight and the other is assigned near minimum weight. For this edge, $f_x(e^4)\, f_y(e^4) \approx B_x\, M_y$.

Finally note that edge e^5 has $x_i \approx 0.5W$ and $y_i \approx -0.5H$. The modulation factors are therefore assigned minimum weights: $f_x(e^5)\, f_y(e^5) \approx B_x\, B_y$.

6.4 Associating the Estimated Interconnect Area with the Cell Edges

In this section, the estimated interconnect area will be associated with cell edges in contrast to channel edges. Note that each internal channel edge is bordered by two cell edges. Hence, it is approximated that the interconnect area associated with a cell edge is one half that computed for a channel edge with the same coordinate pairs (that is, at the same location). Let i now represent a cell edge, and further let e_w^i represent the estimated interconnect area associated with cell edge i. From Eqn. 6.8, it then follows that

$$e_w^i = \frac{1}{2} C_W f_x(x_i) f_y(y_i) \tag{6.13}$$

where the values of x_i and y_i represent the effective positions of cell edge i, computed in the manner of Eqns. 6.11 and 6.12.

By the analysis in Section 6.2, the *desired* value of e_w^i is $0.5C_W$. Hence, a constant α is used to normalize the product of the two rightmost terms in Eqn. 6.13. That is,

$$e_w^i = \frac{1}{2} \alpha\, C_W f_x(x_i) f_y(y_i) \tag{6.14}$$

The value of α is computed as follows:

$$\alpha = \frac{1}{HW} \int_{-0.5H}^{0.5H} \int_{-0.5W}^{0.5W} f_x(x) f_y(y) \; dx\, dy$$

Note that the function $f_x()$ does not depend on y and that the function $f_y()$ does not depend on x. Also, suppose that

$$M = M_x = M_y \tag{6.15}$$

and

$$B = B_x = B_y \tag{6.16}$$

Then with the use of symmetry,

$$\sqrt{\alpha} = \frac{1}{W} \int_{-0.5W}^{0.5W} f_x(x) \; dx \tag{6.17}$$

The use of Eqn. 6.6 leads to:

$$\sqrt{\alpha} = \frac{1}{W} \int_{-0.5W}^{0.5W} M - \frac{(M-B)}{0.5W} \left| x \right| \; dx \tag{6.18}$$

Simplification proceeds as follows:

$$\sqrt{\alpha} = \frac{1}{W} \int_{-0.5W}^{0.5W} M\,dx + \int_{-0.5W}^{0} \frac{(M-B)}{0.5W^2} x\,dx - \int_{0}^{0.5W} \frac{(M-B)}{0.5W^2} x\,dx \tag{6.19}$$

$$\sqrt{\alpha} = M - \frac{(M-B)}{0.25W^2} \int_{0}^{0.5W} x\;dx \tag{6.20}$$

$$\sqrt{\alpha} = M - \left[\frac{(M - B)}{0.25W^2} \right] \frac{W^2}{8} \tag{6.21}$$

and finally,

$$\alpha = \left[\frac{M + B}{2} \right]^2 \tag{6.22}$$

The expected interconnect area associated with cell edge i is given by Equation 6.14, in conjunction with Eqns. 6.11 and 6.12.

6.5 Interconnect-Area Estimation as a Function of Relative Pin Density

The interconnect area associated with a cell edge i is due to two factors: (1) the routing of nets which do not have pins on edge i (that is, routes which simply use the channel bordered by edge i), and (2) the routing of nets with pins on cell edge i. This section presents the portion of the model which accounts for modulation of the baseline average channel width due to the latter factor.

The *pin density* of a cell edge i, represented by d_p^i, is defined as the total number of pins belonging to edge i divided by the length of edge i. The *average pin density* for the entire circuit is represented by D_p. This is computed by dividing the total number of pins (for all of the cells) by the sum of the perimeters of all the cells. It is then possible to define the *relative pin density* of a cell edge i as d_{rp}^i, where

$$d_{rp}^i = \frac{d_p^i}{D_p} \tag{6.23}$$

The interconnect area modulation factor associated with the relative pin density of cell edge i has the following form:

$$f_3(i) = \max \left\{ 1.0, d_{rp}^i \right\} \tag{6.24}$$

Note that $f_3(i) \geq 1.0$ for all i. This was done so that at least some interconnect area is assigned to a cell edge even if the edge has relatively few (if any) pins.

6.6 The Implementation of the Dynamic Interconnect–Area Estimator

The complete expression for the interconnect area associated with a cell edge i is based on the results of the previous two sections. Combining Eqns. 6.14 and 6.24 yields:

$$e_w^i = \frac{1}{2} \alpha \, C_W \, f_x(x_i) \, f_y(y_i) \, f_3(i) \tag{6.25}$$

where α is given by Eqn. 6.22, and where the values x_i and y_i are the effective positions of cell edge i, computed in the manner of Eqns. 6.11 and 6.12.

In the simulated annealing approach used in TimberWolfMC, sufficient interconnect space between the cells is maintained by appending a border around the contours of each cell. The area occupied by each rectilinear cell is represented as a set of one or more nonoverlapping rectangular tiles. Each time the amount of overlap (if any) is updated for two cells, the edges of each tile for a cell are expanded outward. That is, the left edge i_l of each tile is extended farther to the left by an amount $e_w^{i_l}$, given by Eqn. 6.25. Similarly, the right edge i_r of each tile is extended farther to the right by an amount $e_w^{i_r}$, computed in the manner of Eqn. 6.25. Each top tile edge, i_t, is extended farther upward by an amount $e_w^{i_t}$ and each bottom tile edge, i_b, is extended farther downward by an amount $e_w^{i_b}$, similarly obtained from Eqn. 6.25.

The respective outward expansions of the tile edges are updated each time a cell participates in the generation of a new state in the simulated annealing approach. Hence the estimation of the interconnect area attributable to a given edge of a cell is a dynamic quantity, depending on the location of the edge at any given time. It follows that if a cell is moved from a corner of the core area towards the center of the core area, then the effective area of the cell will increase. Similarly, if a cell is moved from the center towards a corner of the core area, then the effective area of the cell will decrease.

The computation time necessary to update the estimated interconnect area for each new state (that is, for each time a cell is moved) is negligible. From Eqn. 6.25, the term C_W, given by Eqn. 6.5, and the term $f_3(i)$ for all i, given by Eqn. 6.24, are computed and stored at the beginning of the execution of the algorithm. Factors $f_x(x_i)$ and $f_y(y_i)$ must be computed for each *moved* edge for each new state. Arguments x_i and y_i are the effective positions of the (moved) cell edge i, computed from Eqns. 6.11 and 6.12. This computation requires approximately three *compare* instructions for each of the two arguments. The substitution of these arguments and the evaluation of Eqns. 6.6 and 6.7 essentially require one multiplication and one subtraction per equation. Hence, the computation time imposed by the dynamic interconnect area estimator does not have an observable effect on the running time of the TimberWolfMC program.

6.7 Results

As described in Chapter 4, the TimberWolfMC placement algorithm proceeds in two distinct stages. During the first stage, which uses a simulated annealing algorithm, the interconnect area around the individual cells is determined using the dynamic interconnect–area estimator. Following the minimization of the total estimated interconnect length, the algorithm proceeds to the second stage. During this stage, three iterations of a placement–refinement algorithm are performed. Even if the overall interconnect–area estimation was accurate, it is quite likely that it was not perfect for each channel. Hence an iterative process was developed to adjust the routing area. By iteratively applying a placement–refinement algorithm, better global results would be expected. Each iteration consists of: (1) a channel definition step, (2) a global routing step, and (3) a placement–refinement step. This global routing information obtained after (2) above is used to compute the density of all of the channels, which then permits accurate determination of the interconnect area. The placement of the cells is then refined in (3) above, this time using a low–temperature simulated annealing algorithm, taking into account the new interconnect areas.

Based on observations of numerous (single–level metal) industrial circuits at Gould–AMI, Intel, and Advanced Micro Devices, it appeared that channels in the very center of a chip were typically on the order of four times wider than those at the very corners of the chip. As a consequence, the dynamic interconnect–area estimator has been used with the parameters M and B in Eqn. 6.22 set to 2.0 and 1.0, respectively. With this selection of parameters, $\alpha = 2.25$. Note that a technology change which leads to some other relationship between the widths of channels at the center of the chip compared to those at the fringes is readily accommodated by an appropriate selection of parameters M and B. This selection process can be guided by the results of the Stage 2 placement–refinement algorithm.

In order to determine the accuracy of the dynamic interconnect–area estimator, the final total interconnect length and the final core area for several circuits were compared at the end of each of the two stages. A large change in the final total interconnect length or core area would indicate substantial cell movement during the second stage of TimberWolfMC, thereby implying inaccurate interconnect area estimation. The results for nine industrial circuits are shown in Table 6.1.

Circuit	Cells/Pads	Nets	Pins	Trials	Percent Interconnect Reduction	Percent Area Reduction
i1	33/38	121	452	5	5.8	3.0
p1	11/45	83	309	6	2.0	–9.2
x1	10/2	267	762	4	4.0	2.5
i2	23/17	127	577	5	–1.0	–3.8
i3	18/10	38	102	2	10.5	–0.5
l1	62/0	570	4309	4	2.5	–0.5
d2	20/139	656	1776	4	12.7	8.5
d1	17/0	288	837	4	0.5	8.25
d3	17/5	136	665	2	0.5	–1.0
avg.					4.4	4.1

Table 6.1 The comparison of total interconnect length and core area at the end of stage 2 versus stage 1 of TimberWolfMC demonstrates the accuracy of the dynamic estimator.

The last two columns represent, respectively, the total interconnect length and the core area at the end of stage 2 divided by the values obtained at the end of stage 1, expressed in terms of a percentage change.

For the nine circuits tested, on average, there was negligible change in core area between the two stages of TimberWolfMC. Furthermore, there was a minimal decrease in total interconnect length after stage 2.

The accuracy of the dynamic interconnect–area estimator is further demonstrated by Figures 6.3–6.6, in which the core–enclosing rectangle is the same size for all of the figures. Figure 6.3 shows the expanded cells for circuit i1 at the end of an execution of stage 1 of TimberWolfMC. Figure 6.4 shows the expanded cells for the same circuit after the conclusion of stage 2. Figures 6.5 and 6.6 present the actual cell placement at the end of stage 1 (Figure 6.5) and after the end of stage 2 (Figure 6.6). For this particular example, the total interconnect length at the end of stage 2 was only 6 percent less than that at the end of stage 1. Furthermore, the core area was only 5 percent less at the end of stage 2 compared to that at the end of stage 1.

This dynamic estimator has enabled the generation of placements which require very little placement modification during or after detailed routing. The estimator is an integral part of the TimberWolfMC macro/custom cell placement and global routing package.

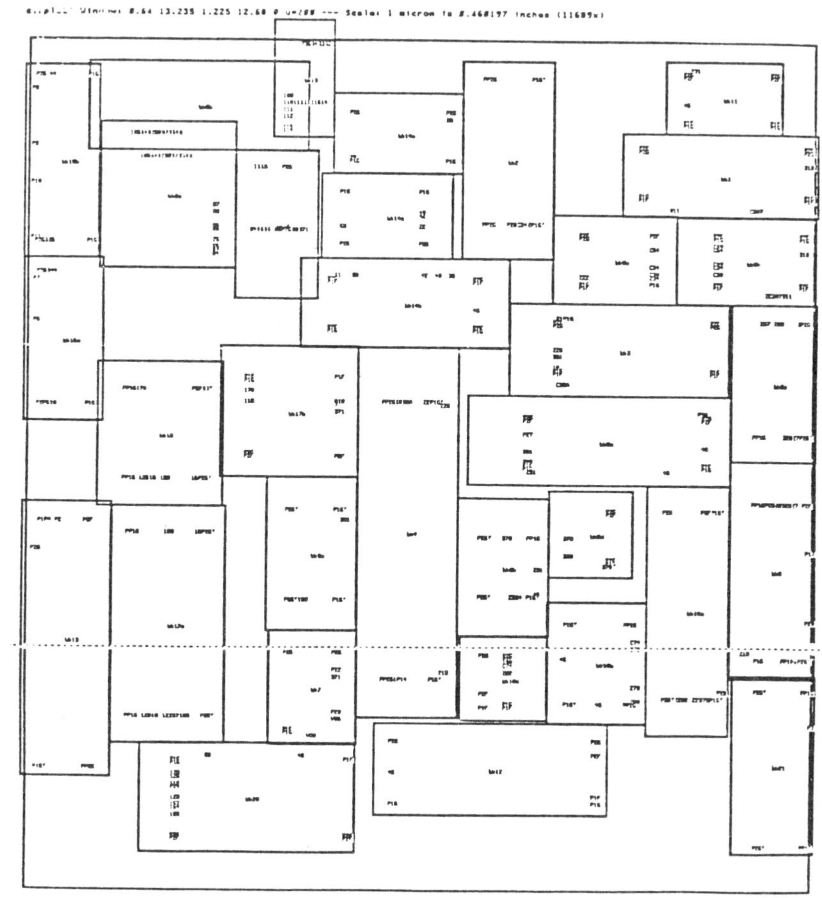

Figure 6.3 The expanded cells for circuit i1 at the end of an execution of stage 1 of TimberWolfMC.

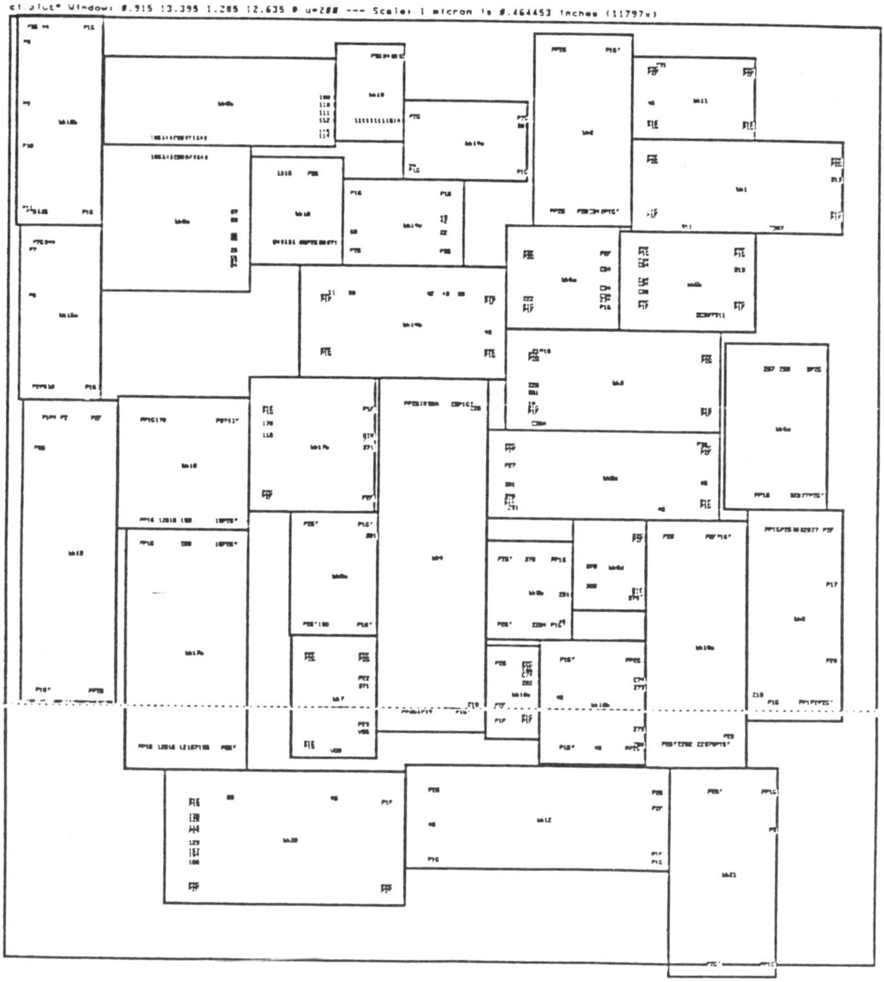

Figure 6.4 The expanded cells for circuit i1 after the conclusion of stage 2 of TimberWolfMC.

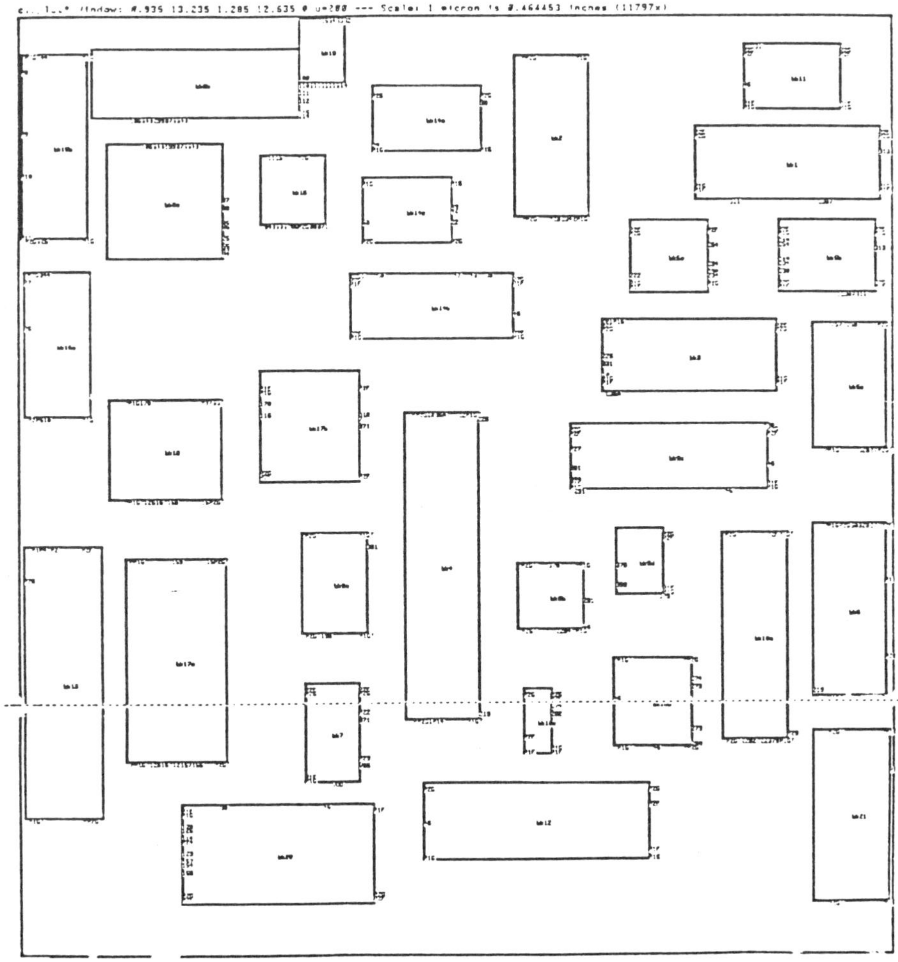

Figure 6.5 The actual cell placement at the end of stage 1 of TimberWolfMC.

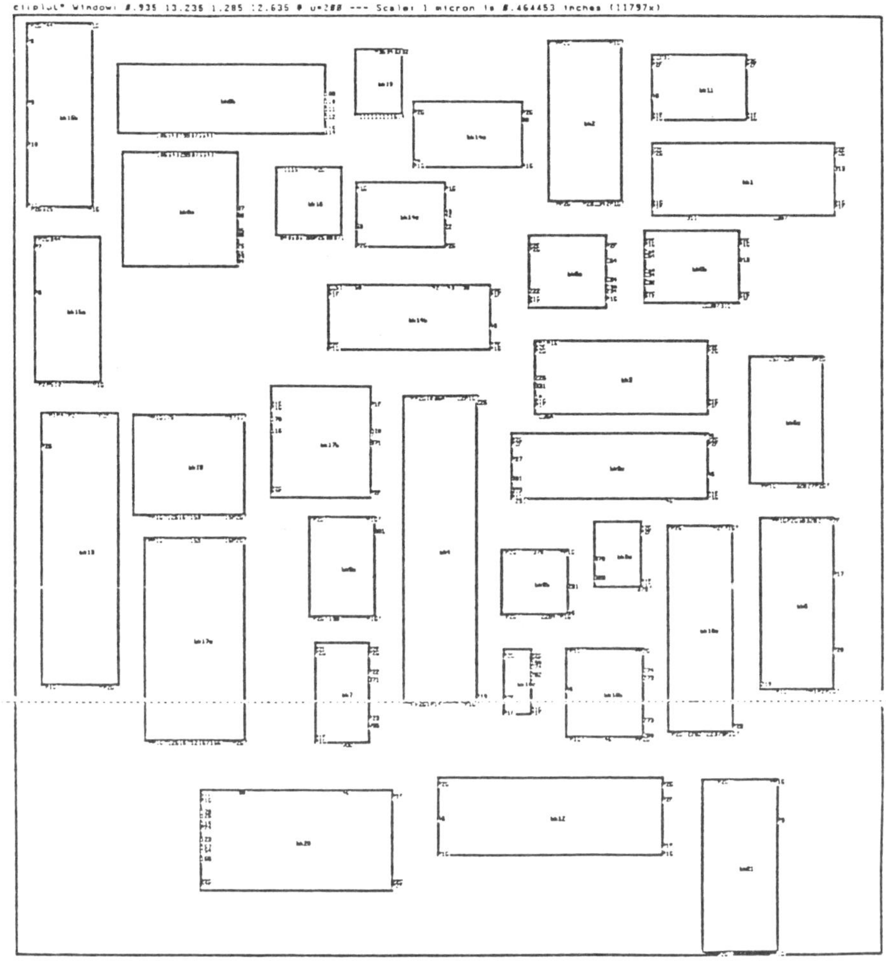

Figure 6.6 The actual cell placement after the end of stage 2 of TimberWolfMC.

Chapter 7

An Edge-Based Channel Definition Algorithm for Rectilinear Cells

7.1 Introduction

This chapter will describe an edge-based channel definition algorithm for circuits consisting of cells of any rectilinear shape. This algorithm is particularly well suited for channel definition just prior to placement refinement. Furthermore, the algorithm may be employed as a basic channel definer just prior to the execution of global and then detailed routers.

As described in Chapter 4, TimberWolfMC begins with a simulated annealing based placement stage, in which the required interconnect area is estimated (using the dynamic interconnect-area estimator described in the previous chapter). After this stage, four basic steps remain in order to complete an automatic layout of a macro/custom cell circuit. (1) A placement-refinement step is performed in which any (usually small) inaccuracies produced by the dynamic interconnect-area estimator are corrected. That is, if insufficient space was allocated between a pair of cells, then additional space is provided as required. Or, if excessive space was allocated, then the cells are compacted as much as possible. (2) The areas (channels) available for routing are identified. This is known as the *channel definition* step. (3) The nets (as specified by the net list) are *globally* routed. By taking a global view of the placement of the interconnections, the global router assigns each portion (segment) of a net to a specific channel. (4) Interconnect instances are created for each of the net segments in each channel. That is, a *detailed router* is used to give each net segment a specific size, shape and location.

In order to perform the placement-refinement step (step 1 above), it is necessary to use steps 2 and 3. The same step 3 global router algorithm for assigning the net segments to specific channels may be used in step 1. On the other hand, a channel definition algorithm applicable to step 2 may be unsuitable for use in step 1. As an example, consider the routing channel defined by the shaded region in Figure 7.1.

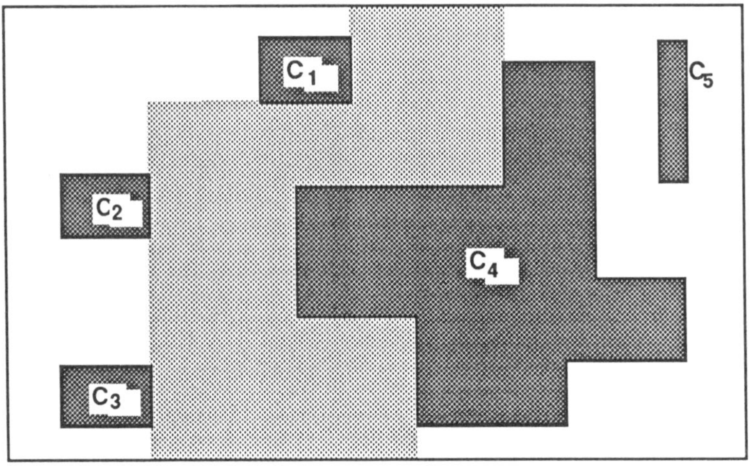

Figure 7.1 The shaded region is an example of a channel defined by traditional channel definers in a rectilinear cell environment.

Suppose that a global router assigned some number of net segments to this channel, which is understood to follow the contour from *a* to *h* as shown in Figure 7.2.

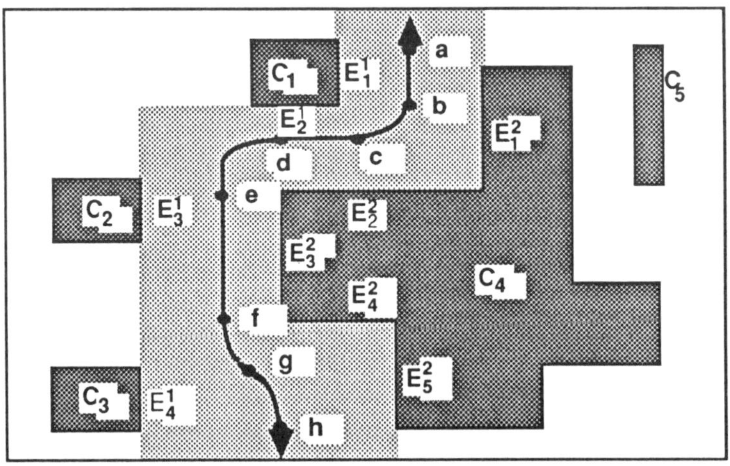

Figure 7.2 A generalized channel having a contour from *a* to *h*. Defining the channel density is a very difficult problem since the sides of the channel are not parallel.

With respect to this channel, the following observations are made: (1) Compaction (or expansion) of the channel is performed by moving the group of cells G_1 independently from the group of cells G_2, where

$$G_1 = \{ c_1, c_2, c_3 \} \tag{7.1}$$

and

$$G_2 = \{ c_4, c_5 \} \tag{7.2}$$

(2) The cell edges bordering the channel shown in Figure 7.2 are labeled E_j^i, where i is the group (either 1 or 2), and where j identifies a particular cell edge in the group. Note that there are four cell edges bordering the channel which belong to group 1 and five cell edges bordering the channel which belong to group 2. (3) Note that defining the *channel density* (the usual congestion metric) is a very difficult problem since the sides of the channel are not parallel. Hence it is not possible to specify a single parameter which gives the expected width (or height) of the channel.

A channel such as that of Figure 7.2 is suitable for detailed routing. However, several disadvantages arise if such a channel is present during the placement-refinement phase. First, since there is no single parameter expressing a dimension of the channel, this channel must be analyzed in greater detail. That is, a multitude of parameters must be derived which specify the required spacing between edges: E_1^1 and E_1^2, E_2^1 and E_2^2, E_3^1 and E_3^2, etc. Second, if this additional analysis reveals that the spacing between cell edges E_1^1 and E_1^2 requires adjustment (either expansion or compaction), such a change may or may not subsequently necessitate an adjustment of the spacing between edges E_3^1 and E_3^2 and perhaps between edges E_4^1 and E_5^2 as well.

In summary, complex channels such as that represented in Figure 7.2 require additional *local* congestion analysis, well beyond a simple channel density calculation. Furthermore, responses to spacing changes may well trigger additional local analyses and/or spacing adjustments.

On the other hand, if a channel is bordered by only two cell edges, then it is possible to specify a single parameter which yields the expected width, or height, of a channel. Defining the *track spacing* (t_s) to be the minimum center-to-center spacing of the interconnect along the length of the channel, the expected width w (for two layers of interconnect) is given by:

$$w = t_s\,(d + 2) \tag{7.3}$$

where d is the density of the channel. Eqn. 7.3 is based on the fact that channel routers are currently available which routinely route a channel in a number of tracks t such that $t \le (d + 1)$.[1] The expected width of a channel is then simply $w = (t + 1)\, t_s$.

It is therefore possible to specify the required spacing between any two cell edges provided that every routing channel is bordered by exactly two cell edges, or bordered by one cell edge and the border of the chip. This fact led to the development of the channel definition algorithm described in this chapter, which may be used either in the placement-refinement phase or the pre-detailed routing phase of the automatic layout procedure. When used immediately prior to detailed routing, the channel definer generates detailed-routing channels such that a guaranteed noncyclic routing order is produced, with one con-

[1] J. Reed, A. Sangiovanni-Vincentelli, and M. Santamauro, "A New Symbolic Channel Router: YACR2," (1985).

dition.[1] This condition assumes that a two-dimensional, or switch-box, router is available, that is, a router which can handle channels of $n \geq 4$ sides, with fixed pins on any of the n sides. Several two-dimensional routers have been reported.[2] [3] [4] [5] [6] [7] [8]

The most fundamental requirement in placement refinement is that each channel must be bordered by exactly two cell edges, or by one cell edge and one chip edge. One exception, however, is permitted. A set of channels which are parallel, adjacent, and equal in length are allowed since the width of each channel can be determined independently.

Several channel definition algorithms have been reported previously.[9] [10] [11] [12] [13] [14] [15] [16] To the best of my knowledge, all have been intended for pre-detailed routing. Unfortunately, for the placement-refinement phase, none of these algorithms is directly applicable. For example, an *empty room,* that is, an area which is neither a cell, a channel, or a node (an intersection of channels) cannot be tolerated.[17] With empty rooms, it is no longer true that a channel is bordered by two cell edges, or a cell edge and a chip edge. Also, the pure slicing structure is not applicable since a channel will be typically bordered by many cell edges.[18] Furthermore, it is likely that a given cell placement will not have the slicing structure. With respect to the placement-refinement phase of

---

[1] A. Sangiovanni-Vincentelli, Private Communication, (1986).

[2] C. P. Hsu, "A New Two-Dimensional Routing Algorithm," (1982).

[3] W. K. Luk, "A Greedy Switch-Box Router," (1984).

[4] M. Marek-Sadowska, "Two-Dimensional Router for Double Layer Layout," (1985).

[5] R. Joobbani and D. Siewiorek, "WEAVER: A Knowledge-Based Routing Expert," (1985).

[6] R. Joobbani, "An Artificial Intelligence Approach to VLSI Routing," (1986).

[7] G. Hamachi and J. Ousterhout, "A Switchbox Router with Obstacle Avoidance," (1984).

[8] H. Shin and A. Sangiovanni-Vincentelli, "A Detailed Router Based on Incremental Routing Modifications: MIGHTY," (1987).

[9] N. P. Chen, C. P. Hsu, E. S. Kuh, C. C. Chen, and M. Takahashi, "BBL: A Building-Block Layout System for Custom Chip IC Design," (1983).

[10] N. P. Chen, C. P. Hsu, and E. S. Kuh, "The Berkeley Building-Block (BBL) Layout System for VLSI Design," (1983).

[11] N. P. Chen, "Routing System for Building Block Layout," (1983).

[12] R. Otten, "Automatic Floorplan Design," (1982).

[13] R. Rivest, "The PI (Placement and Interconnect) System," (1982).

[14] J. Ousterhout, G. Hamachi, R. Mayo, W. Scott, and G. Taylor, "Magic: A VLSI Layout System," (1984).

[15] G. Hamachi and J. Ousterhout, "A Switchbox Router with Obstacle Avoidance," (1984).

[16] W. M. Dai, T. Asano, and E. S. Kuh, "Routing Region Definition and Ordering Scheme for Building Block Layout," (1985).

[17] W. M. Dai, T. Asano, and E. S. Kuh, "Routing Region Definition and Ordering Scheme for Building Block Layout," (1985).

[18] R. Otten, "Automatic Floorplan Design," (1982).

channel definition, several methods suffer from the fact that a channel typically will be bordered by more than two cell edges.[1 2 3]

With respect to the detailed-routing order of the channels, Otten proposed an algorithm which guarantees a noncyclic routing order for rectangular cell placements which have the slicing structure.[4] Dai, Asano, and Kuh were able to achieve a guaranteed noncyclic routing order with the extension to nonslicing placements of rectangular cells by allowing the possibility of L-shaped channels.[5] Recently, Sangiovanni-Vincentelli has shown that it is possible to achieve a guaranteed noncyclic routing order for rectilinear cell shapes by allowing arbitrarily shaped channels.[6] A version of this latter technique was adopted in the algorithm of this chapter. This version required that a channel can have any shape, for example, straight, L, U, rectilinear Z, etc.

Portions of the placement-refinement phase of the channel definition algorithm of this chapter are quite similar to an algorithm which was reported previously.[7 8 9] In fact, the *bottlenecks* defined by N. P. Chen are very similar to the critical regions defined in Section 7.2.[10] The algorithm of this chapter is somewhat more general in that every possible critical region is identified, and used, during the placement-refinement phase. The method by N. P. Chen assumes that two critical regions, or bottlenecks, will not overlap. Overlapping can occur if a critical region is created both by a pair of vertical cell edges and a pair of horizontal cell edges. In the work of N. P. Chen, either the placement is modified to remove the overlapping critical regions, or one of the critical regions is ignored. The ignored critical region is that which is estimated to have the least restriction for the global router, that is, the widest critical region. Since global routing has not been performed yet, this can only be an estimate. In the algorithm of this chapter, all of the critical regions are identified and used.

The remainder of this chapter is divided into four sections. Section 7.2 describes the basic channel definition algorithm. In Section 7.3, the algorithm for generating the channel graph is presented. Next, in Section 7.4, the channel merging algorithm is de-

---

[1] J. Ousterhout, G. Hamachi, R. Mayo, W. Scott, and G. Taylor, "Magic: A VLSI Layout System," (1984).

[2] G. Hamachi and J. Ousterhout, "A Switchbox Router with Obstacle Avoidance," (1984).

[3] R. Rivest, "The PI (Placement and Interconnect) System," (1982).

[4] R. Otten, "Automatic Floorplan Design," (1982).

[5] W. M. Dai, T. Asano, and E. S. Kuh, "Routing Region Definition and Ordering Scheme for Building Block Layout," (1985).

[6] A. Sangiovanni-Vincentelli, Private Communication, (1986).

[7] N. P. Chen, C. P. Hsu, E. S. Kuh, C. C. Chen and M. Takahashi, "BBL: A Building-Block Layout System for Custom Chip IC Design," (1983).

[8] N. P. Chen, C. P. Hsu and E. S. Kuh, "The Berkeley Building-Block (BBL) Layout System for VLSI Design," (1983).

[9] N. P. Chen, "Routing System for Building Block Layout," (1983).

[10] N. P. Chen, "Routing System for Building Block Layout," (1983).

scribed. Finally, Section 7.5 presents the algorithm which prescribes the routing order for the channels, such that no cyclic routing order is produced.

7.2 The Basic Channel Definition Algorithm

This section presents the algorithm for defining the channels for a placement of rectilinear cells in a rectangular region. The channels are generated to have the property such that at most two (parallel) cell edges border the channel. Cell edges which border such a channel will be referred to as *fixed edges*. Further, it is assumed that any fixed pins belonging to a channel reside on these fixed edges. Interconnections entering an end of the channel and subsequently passing through the opposite end of the channel are said to pass along the *length* of the channel. The orthogonal dimension, that is, the dimension limited by as many as two fixed edges, will be referred to as the *width* of the channel.

The following notation will be used to identify a cell edge: $E_{l_v}^{c_i}$ indicates that a fixed edge is the l-th edge belonging to cell c_i, and that this edge is parallel to the vertical direction. Similarly, $E_{m_h}^{c_j}$ indicates that a fixed edge is the m-th edge belonging to cell c_j, and that this edge is parallel to the horizontal direction. In Figure 7.3, the edges are labeled such that the superscript represents the cell number and the subscript is the sequence number of the edge, where edge number one emanates vertically from the leftmost of the lowest vertices.

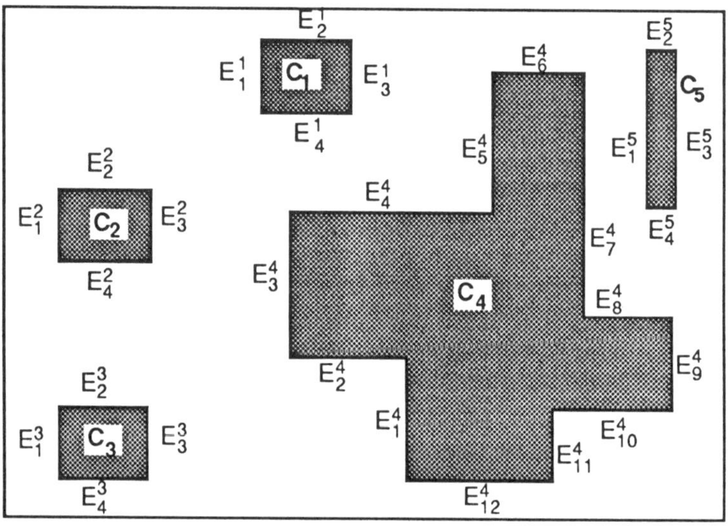

Figure 7.3 The cell edges are labeled such that the superscript represents the cell number and the subscript is the sequence number of the edge, where edge number one emanates vertically from the leftmost of the lowest vertices.

7.2.1 Identifying critical cell-edge pairs

During the placement refinement phase, the critical question is: Given two parallel cell edges belonging to different cells, how close can they be placed such that sufficient interconnect space is made available? Considering first vertical cell edges, it is critical to identify every pair of vertical cell edges, $E_{l_v}^{c_i}$ and $E_{m_v}^{c_j}$, where $i \neq j$ (the edges belong to different cells), such that:

(1v) The projections of the vertical edges $E_{l_v}^{c_i}$ and $E_{m_v}^{c_j}$ on the y-axis have a nonzero overlapping. In Figure 7.4, the projections of the edges E_3^1 and E_1^5 span a common portion of the y-axis equal to the amount f shown in the figure. Also, the projections of the edges E_3^1 and E_5^4 span a common portion of the y-axis equal to the amount g shown in Figure 7.4.

(2v) No other vertical edge $E_{n_v}^{c_k}$ has *both* of the following two properties: (a) it lies between $E_{l_v}^{c_i}$ and $E_{m_v}^{c_j}$ in the x-direction, that is, with $l_{vx} \leq n_{vx} \leq m_{vx}$ or with $m_{vx} \leq n_{vx} \leq l_{vx}$ (where the x subscript implies the x coordinate of the vertical edge) and (b) the projection of the edge on the y-axis overlaps a nonzero portion of the common projection region due to the vertical edges $E_{l_v}^{c_i}$ and $E_{m_v}^{c_j}$.

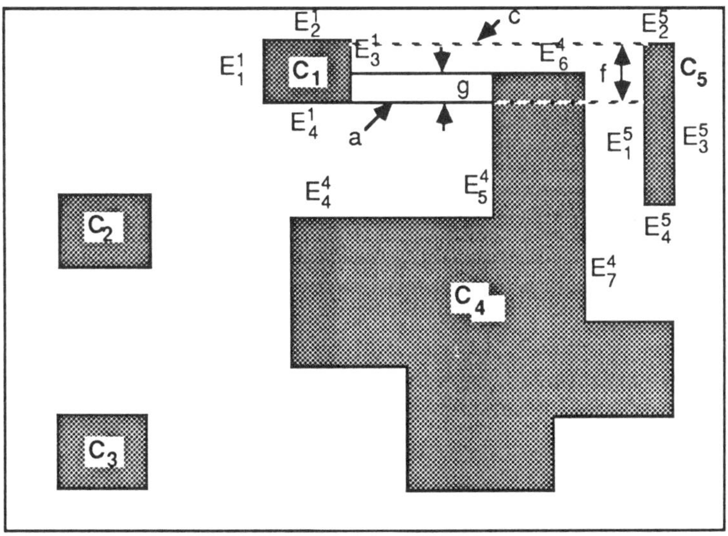

Figure 7.4 The projections of edges E_3^1 and E_1^5 span a common portion of the y-axis equal to the amount f. Also, the projections of edges E_3^1 and E_5^4 span a common portion of the y-axis equal to the amount g.

For example, edges E_3^1 and E_5^4 of Figure 7.4 satisfy both properties (1v) and (2v). However, edges E_3^1 and E_1^5 satisfy only property (1v) since either edge E_5^4 or edge E_7^4 satisfies both properties (2v-a) and (2v-b).

Suppose that the vertical edges $E_{l_v}^{c_i}$ and $E_{m_v}^{c_j}$ satisfy both properties (1v) and (2v) above. Then the projections of these two vertical edges on the y-axis have a common span of $l_c > 0$, where l_c denotes the length of the critical region. The width of the critical region (w_c) is given by:

$$w_c = |l_{vx} - m_{vx}| \quad (7.4)$$

As an example, in Figure 7.4, the edges E_3^1 and E_5^4 generate a *critical region* of length $l_c = g$ and width $w_c = a$. The term *critical* is used for the following reason: Since any fixed pins are located on the sides (parallel to the length l_c) of the critical region, the density of the routing along l_c must be accommodated by the width w_c of the region. Referring to Eqn. 7.3, during placement refinement it is desired to adjust the placement of the cells such that:

$$w_c = w = t_s (d + 2) \quad (7.5)$$

where d is the density of the routing in the critical region (channel).

Similarly, it is critical to identify every horizontal pair of cell edges, $E_{l_h}^{c_i}$ and $E_{m_h}^{c_j}$, where $i \neq j$, such that:

(1h) The projections of the horizontal edges $E_{l_h}^{c_i}$ and $E_{m_h}^{c_j}$ on the x-axis have a nonzero overlapping.

(2h) No other horizontal edge $E_{n_h}^{c_k}$ has *both* of the following two properties: (a) it lies between $E_{l_h}^{c_i}$ and $E_{m_h}^{c_j}$ in the y-direction, that is, with $l_{vy} \leq n_{vy} \leq m_{vy}$ or with $m_{vy} \leq n_{vy} \leq l_{vy}$ and (b) the projection of the edge on the x-axis overlaps a nonzero portion of the common projection region due to the horizontal edges $E_{l_h}^{c_i}$ and $E_{m_h}^{c_j}$.

Suppose that the horizontal edges $E_{l_h}^{c_i}$ and $E_{m_h}^{c_j}$ satisfy both properties (1h) and (2h) above. Then the projections of these two vertical edges on the x-axis have a common span of $l_c > 0$, where l_c denotes the length of the critical region. The width of the critical region (w_c) is given by:

$$w_c = |l_{vy} - m_{vy}| \quad (7.6)$$

7.2.2 Characterization of fixed cell edges

Vertices of a cell are numbered starting from the leftmost of the lowest vertices, and the numbering proceeds in the clockwise fashion. If (in some sense) one were proceeding down the length of a cell edge E from the lower-numbered vertex to the higher-numbered vertex, then the interior of the cell is to the *right* of E. The boundaries of the chip are also treated as *fixed edges*, and since any available routing area would be *inside* the chip boundary, the interior of this *pseudo cell* is outside the boundary. The four boundary vertices are therefore numbered in a counter-clockwise fashion.

A vertical fixed edge is characterized by: (1) an x-coordinate (E_x) indicating the horizontal position of the edge, and (2) two y-coordinates, E_{y2} and E_{y1}, indicating the vertical positions of the endpoints (vertices) of the edge. The coordinate E_{y1} is the y-position of the lower-numbered vertex, and the coordinate E_{y2} is the y-position of the higher-numbered vertex. (3)A direction (E_d) is assigned to an edge such that $E_d = 1$ if $E_{y2} \geq E_{y1}$ (in some sense, a *rising* vertical edge) and such that $E_d = -1$ if $E_{y2} < E_{y1}$ (in some sense, a *falling* vertical edge).

Similarly, horizontal fixed edges are characterized by: (1) a y-coordinate (E_y) indicating the vertical position of the edge, and (2) two x-coordinates, E_{x2} and E_{x1}, indicating the horizontal positions of the endpoints of the edge. (3) A direction (E_d) is assigned to an edge such that $E_d = 1$ if $E_{x2} \geq E_{x1}$ and $E_d = -1$ if $E_{x2} < E_{x1}$.

For any vertex v (located at the junction of two edges, E_v and E_h), one of the edges (E_{in}) is necessarily directed toward v and the other edge (E_{out}) is necessarily directed away from v. A vertical edge is directed toward v if E_{y2} is also the y-position of v. On the other hand, a horizontal edge is directed toward v if E_{x2} is also the x-position of v.

7.2.3 An algorithm for finding critical regions

A cell-edge based algorithm is used to find the critical regions or channels. A critical region will be denoted as *vertical* if the length of the region l_c (as defined in Section 7.2.1) extends in the vertical (or y) direction and the width of the critical region w_c extends in the horizontal direction. On the other hand, a critical region is said to be *horizontal* if l_c extends in the horizontal direction and w_c extends in the vertical direction.

The algorithm for finding the *vertical* critical regions is now given. The four steps outlined below are applied to each vertical fixed edge E_v, having the property $E_d = -1$. Let v_2 represent the top vertex of edge E_v, and let v_1 represent the bottom vertex. Figure 7.5 shows examples of each of the steps.

(v-1) First, along a horizontal line l_{h2} beginning at v_2 and extending to the right, find the nearest vertical edge E_{v2} which has $E_d = 1$ and which either *fully* intersects line l_{h2}, or is such that only its top vertex intersects l_{h2}. (The term *full* intersection will be used to imply that two perpendicular lines actually cross one another, as opposed to one line terminating on the other). Second, along a horizontal line l_{h1} beginning at v_1 and extending to the right, find the nearest vertical edge E_{v3} which has $E_d = 1$ and which either fully intersects line l_{h1}, or is such that only its bottom vertex intersects l_{h1}. If $E_{v2} = E_{v3}$, then let R_1 represent the region bordered on the left by edge E_v, on the right by edge E_{v2}, and bordered on the top by the line l_{h2} and on the bottom by l_{h1}. If no other fixed edges E_{vj} penetrate R_1, then R_1 is a *critical region.* If R_1 is critical, further examination of edge E_v is unnecessary. An example is shown in Figure 7.5a.

(v-2) First, along a horizontal line l_{h2} beginning at v_2 and extending to the right, find the nearest vertical edge E_{v2} which has $E_d = 1$ and which either fully intersects line l_{h2}, or is such that only its top vertex intersects l_{h2}. Second, let the bottom vertex of E_{v2} be represented by v_{r1}. Along a horizontal line l_{h1} beginning at v_{r1} and extending to the left, find the nearest vertical edge E_{v3} which has $E_d = -1$ and which fully intersects line l_{h1}, or is such that only its bottom vertex intersects l_{h1}. If $E_{v3} = E_v$, then let R_2 represent the region bordered on the left by edge E_v, on the right by edge E_{v2}, and bordered on the top by the line l_{h2} and on the bottom by l_{h1}. If no other fixed edges E_{vj} penetrate R_2, then R_2 is a *critical region.* Figure 7.5b shows such an example.

(v-3) First, along a horizontal line l_{h1} beginning at v_1 and extending to the right, find the nearest vertical edge E_{v2} which has $E_d = 1$ and which either fully intersects line l_{h1}, or is such that only its bottom vertex intersects l_{h1}. Second, let the top vertex of E_{v2} be represented by v_{r2}. Along a horizontal line l_{h2} beginning at v_{r2} and extending to the left, find the nearest vertical edge E_{v3} which has $E_d = -1$ and which fully intersects line l_{h2}, or is such that only its top vertex intersects l_{h2}. If $E_{v3} = E_v$, then let R_3 represent the region bordered on the left by edge E_v, on the right by edge E_{v2}, and bordered on the top by the line l_{h2} and on the bottom by l_{h1}. If no other fixed edges E_{vj} penetrate R_3, then R_3 is a *critical region.* An example is shown in Figure 7.5c.

(v-4) Extend horizontal lines (l_t and l_b) to the right starting from vertices v_2 and v_1, respectively. Find each fixed edge E_{vi} which has the property $E_d = 1$ and which is fully contained (in the vertical direction) by lines l_t and l_b. Let l_{h2} be the horizontal line starting from the top vertex of E_{vi} and extending back to the left to edge E_v. Also, let l_{h1} be the horizontal line starting from the bottom vertex of E_{vi} and extending back to the left to edge E_v. Defining R_4 to be the region bordered on the left by edge E_v, on the right by edge E_{vi}, and bordered on the top by the line l_{h2} and on the bottom by l_{h1}. If no other fixed edges E_{vj} penetrate R_4, then R_4 is a *critical region.* An example of such a critical region is shown in Figure 7.5d.

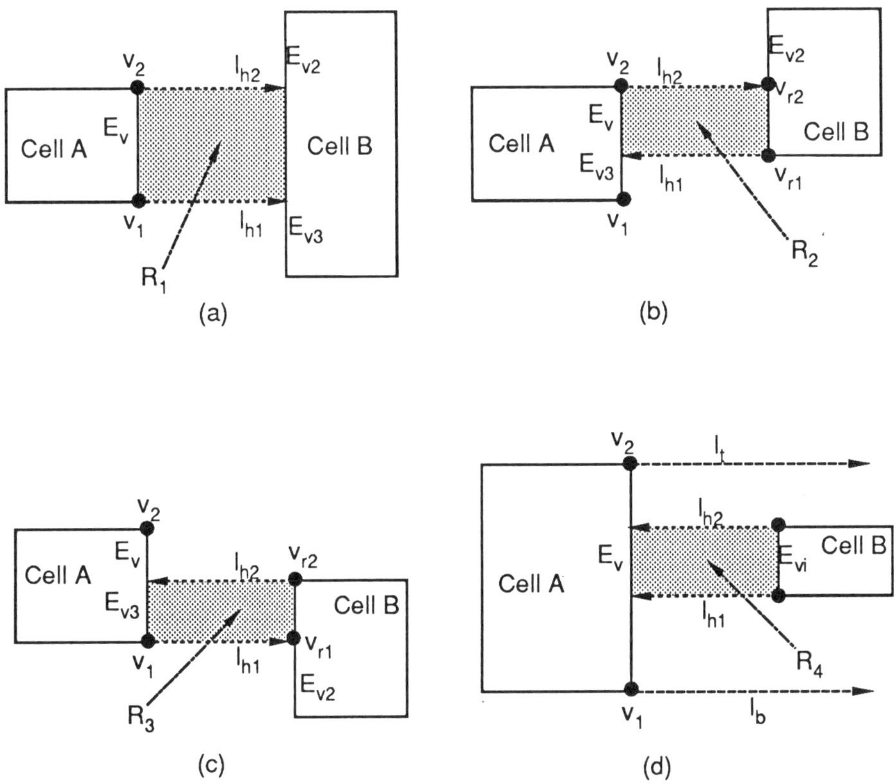

Figure 7.5 (a) Example of step (v-1), where $E_{v2} = E_{v3}$. (b) Example of step (v-2), where $E_v = E_{v3}$. (c) Example of step (v-3), where $E_v = E_{v3}$. (d) Example of step (v-4).

In Figure 7.6, critical region R_1 was generated as a result of step (v-1), R_2 was generated by step (v-2), R_3 was generated by step (v-3) and finally, critical region R_4 was created as a result of step (v-4).

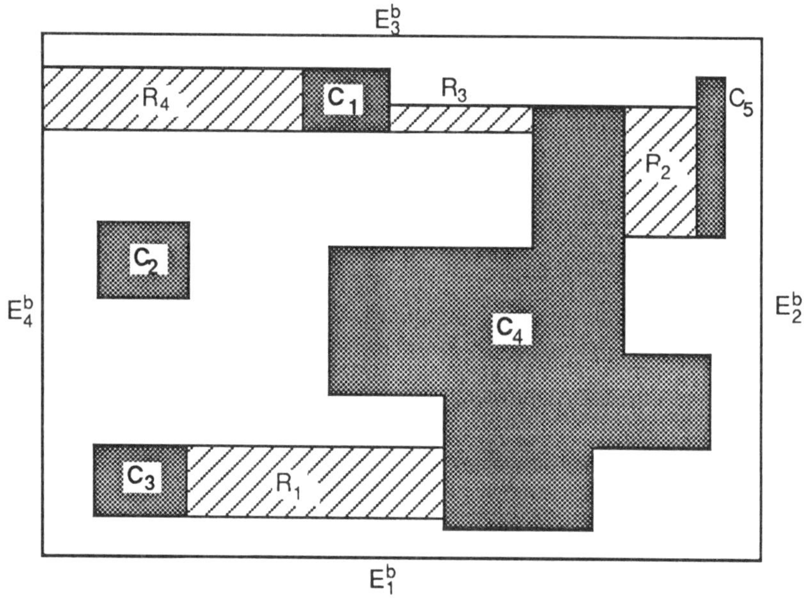

Figure 7.6 Critical region R_1 was generated as a result of step (v-1), R_2 was generated by step (v-2), R_3 was generated by step (v-3) and finally, critical region R_4 was created as a result of step (v-4).

The algorithm for finding the *horizontal* critical regions is similar and will not be presented. The only differences in the description of the algorithm are: (1) The four steps are applied to each horizontal fixed edge E_h, having the property $E_d = 1$. (2) v_2 represents the left vertex of edge E_h, and v_1 represents the right vertex.

Regions, such as R_1, R_2, etc. in Figure 7.5, are termed *critical* because their widths (w_c) are established by a pair of fixed cell edges, E_v and E_{v2} for vertical critical regions and E_h and E_{h2} for horizontal critical regions. Each of these critical regions will become a *channel* in the channel graph used for global routing.

7.3 The Generation of the Channel Graph

The first step in generating the channel graph is to determine all of the critical regions for a placement. As an example, for the placement of Figure 7.3, the application of steps (v-1) through (v-4) for vertical regions and steps (h-1) through (h-4) for horizontal regions have generated the set of critical regions represented by the hatched areas shown in Figure 7.7.

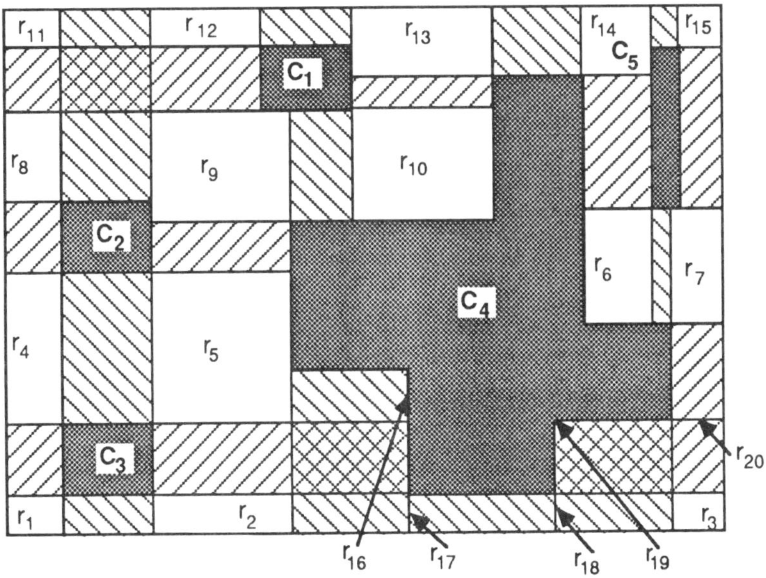

Figure 7.7 For the placement of Figure 7.3, the application of steps (v-1) through (v-4) for vertical regions and steps (h-1) through (h-4) for horizontal regions have generated the set of critical regions represented by the hatched areas.

The nodes of the channel graph are determined based upon the rectilinear regions r_i which are neither part of a cell nor part of a critical region. These regions are referred to as *white-space regions*. In Figure 7.7, the white-space regions are labeled r_1 through r_{15}. Note that white-space regions may have zero area, for example, r_{16} through r_{20}.

The initial location of a node n_i is determined based upon the bounding box of the corresponding white-space region r_i. Let the center of the bounding box be represented by the coordinate pair (n_{ix}, n_{iy}). This coordinate pair represents the initial location for node n_i. In Figure 7.8, the locations of the nodes n_i are shown for the white-space regions r_i.

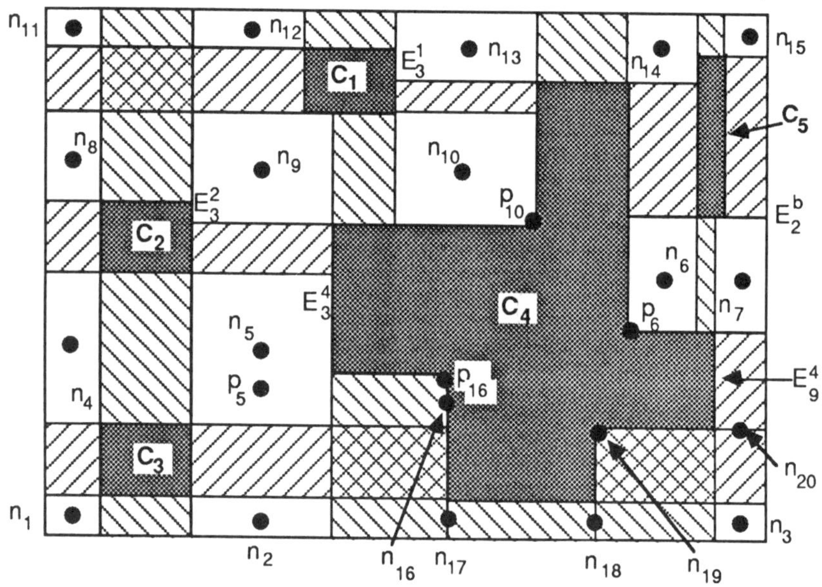

Figure 7.8 The locations of the nodes n_i are shown for the white-space regions r_i.

The construction of the channel graph begins by adding all of the nodes n_i, each of which represents a white-space region r_i. A vertical edge, representing a vertical channel, between nodes n_i and n_j is added to the graph if a vertical critical region borders both white-space regions r_i and r_j. If n_j is above n_i, the vertical critical region must border on the top of r_i and on the bottom of r_j. Similarly, a horizontal edge, representing a horizontal channel, between nodes n_i and n_j is added to the graph if a horizontal critical region borders both white-space regions r_i and r_j. If n_j is to the right of n_i, the horizontal critical region must border on the right of r_i and on the left of r_j.

In Figure 7.8, the vertical channel formed as a result of fixed cell edges E_3^2 and E_3^4 forms an edge in the channel graph with nodes n_5 and n_9. Note that the node pair n_{19} and n_6 does *not* represent a vertical channel. This is because the white-space regions r_{19} and r_6 are not both bordered by a common critical region. However, the node pair n_{20} and n_7 *does* represent a vertical channel. This is because white-space regions r_{20} and r_7 are both bordered by the critical region bounded by fixed edges E_9^4 and E_2^b. Table 7.1 contains the full list of vertical and horizontal channels for the placement in Figure 7.8.

Vertical Edges (Channels)	Horizontal Edges (Channels)
(n_1, n_4)	(n_1, n_2)
(n_4, n_8)	(n_2, n_{17})
(n_{17}, n_{16})	(n_{17}, n_{18})
(n_{18}, n_{19})	(n_{18}, n_3)
(n_8, n_{11})	(n_4, n_5)
(n_2, n_5)	(n_8, n_9)
(n_5, n_9)	(n_9, n_{10})
(n_9, n_{12})	(n_6, n_7)
(n_{10}, n_{13})	(n_{11}, n_{12})
(n_3, n_{20})	(n_{12}, n_{13})
(n_6, n_{14})	(n_{13}, n_{14})
(n_7, n_{15})	(n_{14}, n_{15})
(n_{20}, n_7)	(n_5, n_{16})
	(n_{19}, n_{20})

Table 7.1 Edges (channels) for the channel graph corresponding to the placement in Figure 7.8.

The locations of the nodes are refined as follows: Consider a vertical edge extending from node n_i to node n_j. This edge represents a vertical channel bordered on the left and right sides by fixed cell edges E_l and E_r, respectively, each of which typically has some number of pins. The locations of n_i and n_j are adjusted, if necessary, to ensure that all pin locations lie between the two nodes (in a vertical sense). Let v_t represent the topmost vertex of the top vertices of E_l and E_r. Also, let v_b represent the bottommost vertex of the bottom vertices of E_l and E_r. The position of node n_j is adjusted to equal the vertical position of v_t under the following conditions: If n_j is located below v_t and if either n_j is *not* connected by a vertical edge to a node above itself or if v_t *is* contained by the white-space region r_j. In a similar fashion, the position of node n_i is adjusted to equal the vertical position of v_b under the following conditions: If n_i is located above v_b and if either n_i is *not* connected by a vertical edge to a node below itself or if v_b *is* contained by the white-space region r_i. A corresponding procedure is implemented for horizontal edges.

As an example, in Figure 7.8, the examination of edge (n_5, n_9) will result in n_5 moving to a new location at p_5. Also, node n_{13} is shifted upward slightly to ensure that it lies above the top vertex of edge E_3^1. Furthermore, nodes n_{10}, n_{16}, and n_6 take on the new positions p_{10}, p_{16}, and p_6, respectively, as shown. In Figure 7.9, the channel graph generated for the placement of Figure 7.8 is superimposed on the corresponding cell placement.

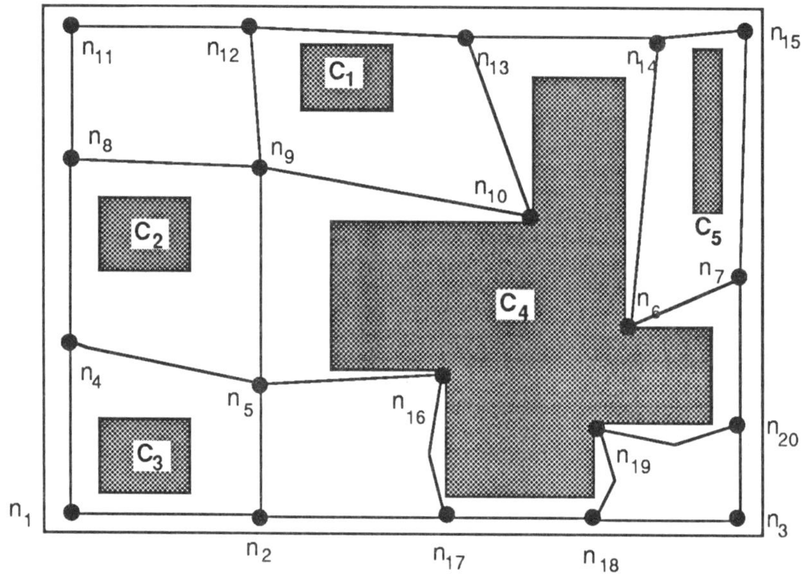

Figure 7.9 The channel graph generated for the placement of Figure 7.8 is superimposed on the corresponding cell placement.

A somewhat pathological placement of seven cells is shown in Figure 7.10. The hatched regions represent the vertical and horizontal critical regions. The positions of the nodes of the channel graph are also shown in the figure. The complete channel graph, containing the edges representing the vertical and horizontal channels is shown in Figure 7.11.

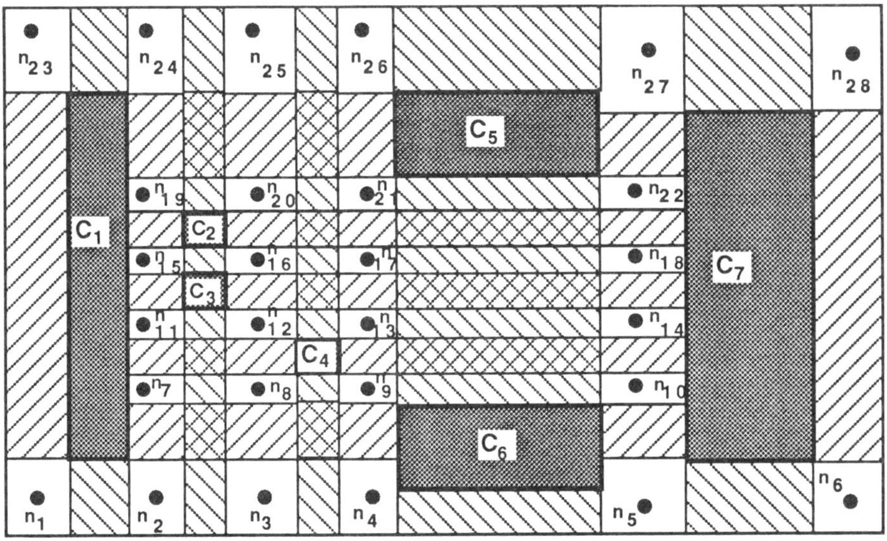

Figure 7.10 A somewhat pathological placement of seven cells. The hatched regions represent the vertical and horizontal critical regions.

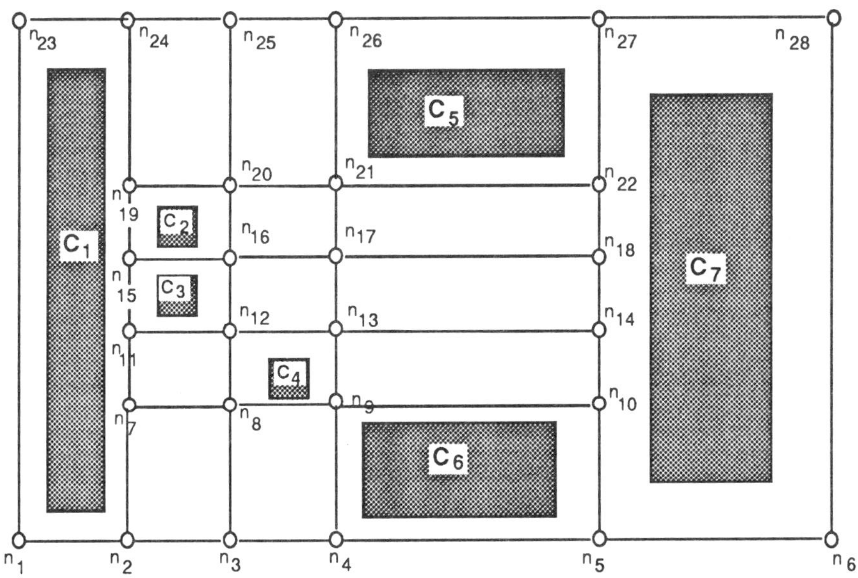

Figure 7.11 The complete channel graph for the placement of Figure 7.10.

Each edge in the channel graph has an associated length and width. The width associated with an edge is given by the width of the critical region represented by the edge.

The length of a vertical edge in the channel graph is given by the vertical separation of the pair of nodes representing the edge. Similarly, the length of a horizontal edge in the channel graph is given by the horizontal separation of the pair of nodes representing the edge.

The channel graph generation algorithm has the virtue that every critical region of the placement is identified and is represented by an edge in the channel graph. This differs from the approach of N. P. Chen, where two critical regions are not allowed to overlap.[1] In his method, either the placement is modified to remove the overlapping critical regions, or one of the critical regions is ignored. The ignored critical region is that which is estimated to have the least restriction for the global router, that is, the widest critical region. Since global routing has not been performed yet, this can only be an estimate. In the algorithm of this chapter, all of the critical regions are identified and used.

7.4 The Generation of the Channel Routing Order

When channel definition is employed just prior to detailed routing, it is necessary to prescribe a *routing order* for the channels. The detailed router is then invoked on each channel in the prescribed order. A guaranteed noncyclic routing order is produced for the set of channels, with one condition.[2] The condition is that a two-dimensional router is available, that is, a router which can handle channels of $n \geq 4$ sides, with fixed pins on any of the n sides.

The algorithm which prescribes the routing order of the channels is a generalization of the *slicing structure*, as first described by Otten, for the arbitrary rectilinear cell case.[3] The basic idea behind slicing is to recursively divide, or slice, a portion of the chip into two parts, starting from the whole chip. As an example, consider the cell placement shown in Figure 7.12. The horizontal line labeled 1 slices the chip into two parts. Next, line 2 slices the top half into two parts. Finally, line 3 slices cells C and D. A cell placement is said to have the slicing structure if the recursive slicing procedure can be applied until each cell is separated.

---

[1] N. P. Chen, "Routing System for Building Block Layout," (1983).

[2] A. Sangiovanni-Vincentelli, Private Communication, (1986).

[3] R. H. Otten, "Automatic Floorplan Design," (1982).

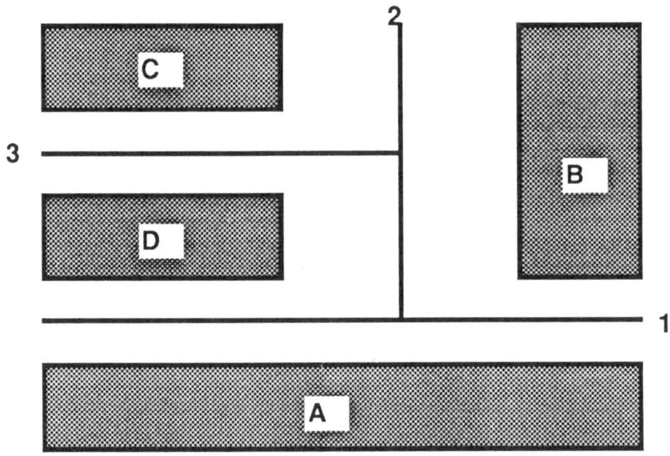

Figure 7.12 The horizontal line labeled 1 slices the chip into two parts. Next, line 2 slices the top half into two parts. Finally, line 3 slices cells *C* and *D*. A cell placement is said to have the slicing structure if the recursive slicing procedure can be applied until each cell is separated.

The slicing scheme has the property that if the channels are numbered in the order they were generated, then a conventional channel router can be applied to detail route the channels in the reverse order of their numbers. A conventional channel router may be used since fixed pins are only present on at most two sides of the channel. Furthermore, these two sides are essentially parallel.

It is apparent that generating the routing order for the channels for cell placements which have the slicing structure is a straight forward procedure. However, not all cell placements have the slicing structure, as can be seen in Figure 7.13. The four *T* junctions create a cyclic precedence relation.[1] [2] The four channels in Figure 7.13 are labeled *a, b, c* and *d*. Note that channel *a* must be routed before *b*, channel *b* before *c*, channel *c* before *d* and finally, channel *d* before *a*, thus leading to a contradiction.

[1] K. Kani, H. Kawanishi, and A. Kishimoto, "ROBIN: A Building Block LSI Routing Program," (1976).

[2] U. Kajitani, "Order of Channels for Safe Routing and Optimal Compaction of Routing Area," (1983).

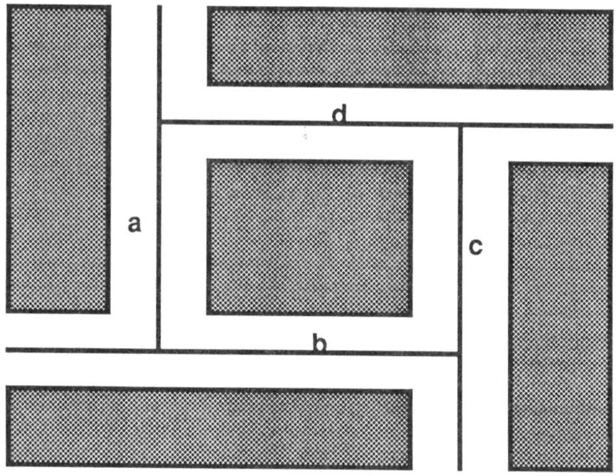

Figure 7.13 Not all cell placements have the slicing structure. The four T junctions create a cyclic precedence relation. The four channels are labeled a, b, c and d. Note that channel a must be routed before b, channel b before c, channel c before d, and finally, channel d before a, thus leading to a contradiction.

A solution to this problem was described by Dai, et. al.[1] In their strategy, L-shaped channels are introduced to break the cyclic precedence relations. In fact, they introduced an algorithm which, for the rectangular cell case, invariably determines a definitive channel routing ordering which minimizes the use of L-shaped channels. In Figure 7.14, the same cell placement as in Figure 7.13 is shown. Note that the introduction of the L-shaped channel labeled 1 cuts the layout into two parts. Conventional channels labeled 2, 3 and 4 are produced along the lines of the usual slicing strategy. As in the slicing structure case, the channels are routed in the opposite order in which they were generated.

While the algorithm by Dai, et. al., is adequate for the rectangular-cell case, it cannot define a routing order for all cases involving general rectilinear cells. One such example is shown in Figure 7.15. Since TimberWolfMC permits cells to have any rectilinear shape, a new ordering scheme was required for detailed routing.

[1] W. M. Dai, T. Asano, and E. S. Kuh, "Routing Region Definition and Ordering Scheme for Building Block Layout," (1985).

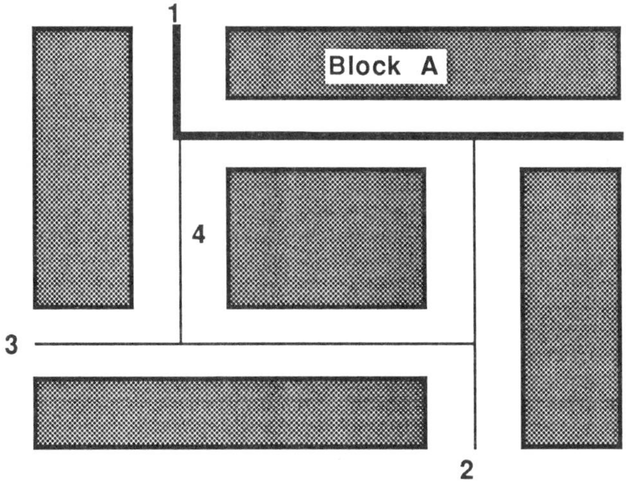

Figure 7.14 The introduction of the *L*-shaped channel labeled 1 cuts the layout into two parts. Conventional channels labeled 2, 3 and 4 are produced along the lines of the usual slicing strategy.

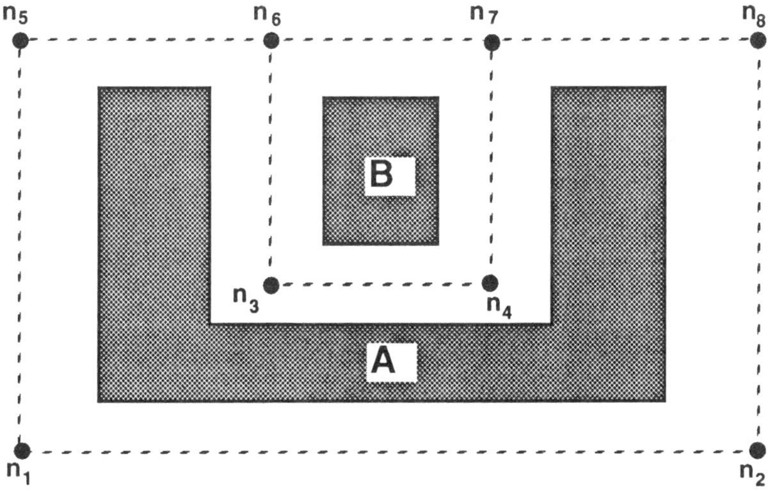

Figure 7.15 The algorithm by Dai, et. al., cannot define a routing order for the example shown here.

The algorithm described in this section handles the general rectilinear cell case. This algorithm assumes the availability of a generalized channel router, that is, a router which can handle channels of $n \geq 4$ sides, with fixed pins on any of the n sides. In fact, such a router was recently described.[1]

The ordering algorithm begins by assigning costs to each edge in the channel graph. The cost of traversing a channel edge (n_i, n_j) in the channel graph is given by $c(n_i, n_j)$:

$$c(n_i, n_j) = -l(n_i, n_j) \tag{7.7}$$

where the expression $l(n_i, n_j)$ represents the length of edge (n_i, n_j) in the channel graph.

The notation P_i is used to denote a path, or sequence of edges, in the channel graph which cuts a portion (or all) of a placement into two parts. P_i will constitute a generalized detailed-routing channel. That is,

$$P_i = n_{i1}, n_{i2}, \cdots, n_{ik} \tag{7.8}$$

hence indicating that k nodes and $k-1$ channel edges constitute the cut. If an edge (n_p, n_q) has been selected as part of a cut of the channel graph, the cost of traversing the edge is given by: $c(n_p, n_q) \to \infty$. That is, if $(n_p, n_q) \in P_i$, then any subsequent path P_j is not permitted to use edge (n_p, n_q).

Summarizing, an edge (n_p, n_q) is assigned a cost $c(n_p, n_q) = -l(n_p, n_q)$ if the edge has *not* been selected as a component of a generalized detailed-routing channel P_i. On the other hand, if the edge *has* already been selected as a component of a generalized detailed-routing channel, then subsequent paths are not permitted to use this edge. This latter condition is implemented by assigning the cost $c(n_p, n_q) \to \infty$ to the edge.

The cost of a path P_i of the form of Eqn. 7.8 is given by the sum of the costs for each edge, plus a penalty each time the path makes an *orthogonal* turn. That is, a horizontal edge either precedes or follows a vertical edge in the path. If a cell placement has the slicing structure, then paths with orthogonal turns are not needed. However, cell placements which do not have a slicing structure require one or more paths with orthogonal turns. For example, in Figure 7.15, the path (n_6, n_3, n_4, n_7) has two orthogonal turns.

The penalty for such orthogonal turns is given by κ, where κ is a positive number larger than the length of any path in the channel graph. Such a selection for κ ensures that, at all times, a path will be selected which has the fewest possible orthogonal turns, regardless of the length of the path.

The expression for the cost of a generalized detailed-routing channel, or path, is now developed. Let P_i constitute such a path, and suppose without loss of generality that this path is expressed as

$$P_i = n_1, n_2, \cdots, n_k \tag{7.9}$$

where k is the number of nodes in the path. Then the expression for the cost of the path may be given as

$$c(P_i) = \sum_{i=2}^{k} \left[c(n_{i-1}, n_i) + p(i-2, i-1, i)\,\kappa \right] \tag{7.10}$$

[1] H. Shin and A. Sangiovanni-Vincentelli, "A Detailed Router Based on Incremental Routing Modifications: MIGHTY," 1987.

The function $p(i-2, i-1, i)$ has value zero if the two edges (n_{i-2}, n_{i-1}) and (n_{i-1}, n_i) are both vertical, or both horizontal. Otherwise, the function $p(i-2, i-1, i)$ has value one. In summary,

$$p(i-2, i-1, i) = \begin{cases} 0 & \text{if } (n_{i-2}, n_{i-1}, n_{i-1}, n_i) \text{ are both vert. or both horiz.} \\ 1 & \text{otherwise} \end{cases} \tag{7.11}$$

The steps of the ordering algorithm will be given now. I will use the placement of six macro cells in Figure 7.16 to illustrate the operation of the various steps. The channel graph for the placement of Figure 7.16 is shown in Figure 7.17.

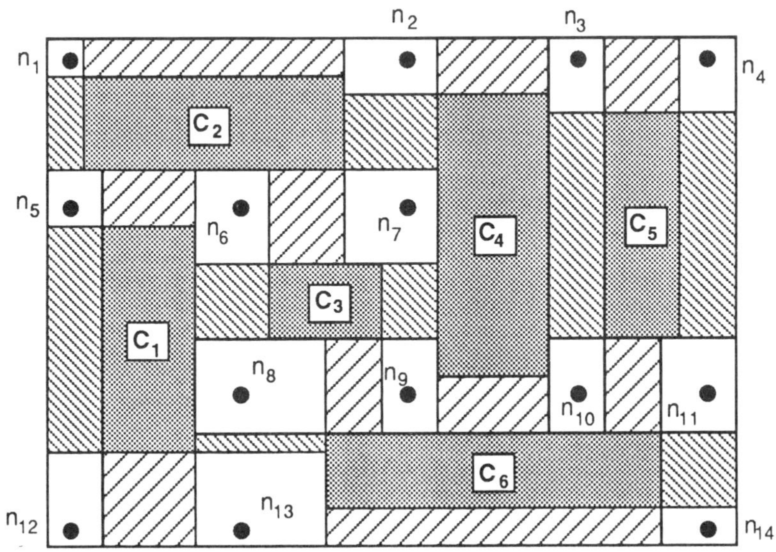

Figure 7.16 The placement of six macro cells is shown, along with the corresponding vertical and horizontal critical regions.

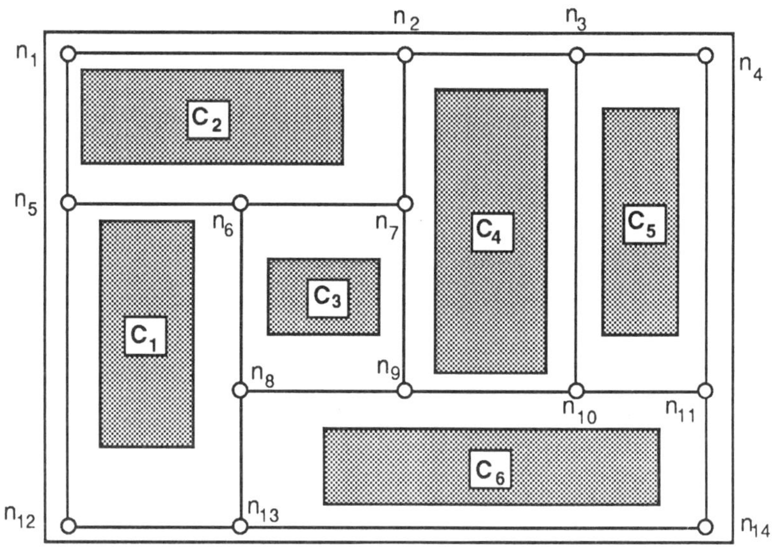

Figure 7.17 The channel graph corresponding to the placement of Figure 7.16.

(1) All *peripheral nodes* are added to the set S. A node n_i is considered on the periphery if it belongs to a channel edge (n_i, n_j) such that this channel is on an outside border (periphery) of the chip. After this step, for the example of Figure 7.17, the set S is given by: $S = \{n_1, n_2, n_3, n_4, n_{11}, n_{14}, n_{13}, n_{12}, n_5\}$. In Figure 7.18, the dark nodes constitute the set S. Set the counter variable: $\gamma = 0$. The selected paths (channels) are $P_1 = (n_1, n_2, n_3, n_4)$, $P_2 = (n_{12}, n_{13}, n_{14})$, $P_3 = (n_1, n_5, n_{12})$, $P_4 = (n_4, n_{11}, n_{14})$. These four paths will be the final channels to be routed.

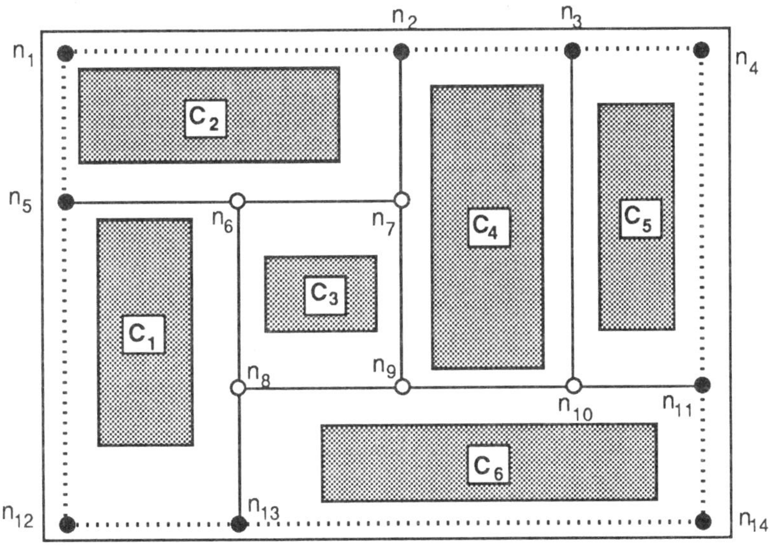

Figure 7.18 The dark nodes belong to the set S for $\gamma = 0$.

(2) Find the lowest cost path between any two nodes n_p and n_q in the set S. That is, for all pairs $(n_i, n_j) \in S$, find the pair of nodes (n_p, n_q) such that Eqn. 7.10 has minimum value. Let the selected path be represented by P_{pq}. Note that since the cost of traversing each *available* edge (Eqn. 7.7) is minus the length of the edge, this step attempts to find the *longest path* between any two nodes in the set S. Further, by the selection of κ, the selected path will have the fewest orthogonal turns amongst all available paths.

Set $\gamma = \gamma + 1$ and then label path P_{pq} with the value of γ: P^{γ}_{pq}.

(3) For each edge $(n_i, n_j) \in P^{\gamma}_{pq}$, set $c(n_i, n_j) \rightarrow \infty$. Further, add the two nodes for each such edge to the set S (excepting nodes n_p and n_q, which are already in S).

(4) While any edges (n_i, n_j) remain having cost $c(n_i, n_j) < \infty$, go back to step 2.

The generalized detailed-routing channels, expressed by P^{γ}_{pq} above, are then routed in the reverse order of the value of γ. As an example, consider again the cell placement of Figure 7.18. The step for $\gamma = 1$ is shown in Figure 7.19. The path generated is $P_5 = (n_{13}, n_8, n_9, n_{10}, n_{11})$. The augmented set S is shown as dark nodes.

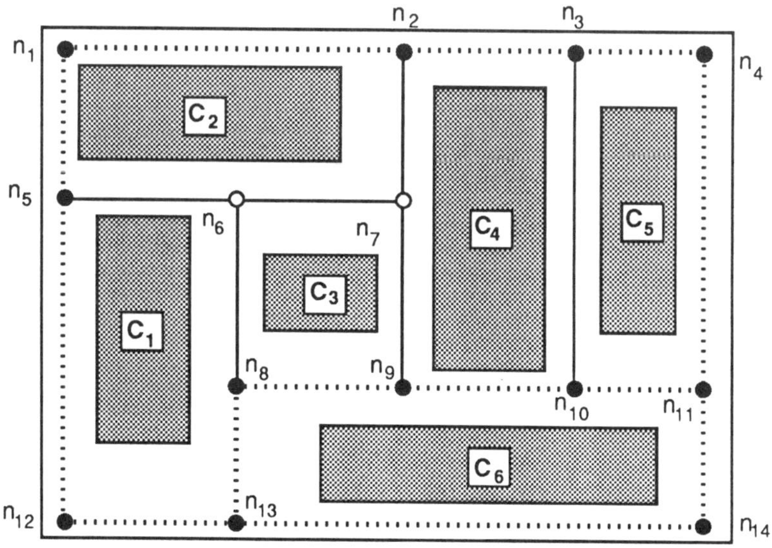

Figure 7.19 The step for $\gamma = 1$.

The step for $\gamma = 2$ is shown in Figure 7.20. The path generated is $P_6 = (n_3, n_{10})$. The set S is shown as dark nodes.

The step for $\gamma = 3$ is shown in Figure 7.21. The path generated is $P_7 = (n_2, n_7, n_9)$. The augmented set S is shown as dark nodes.

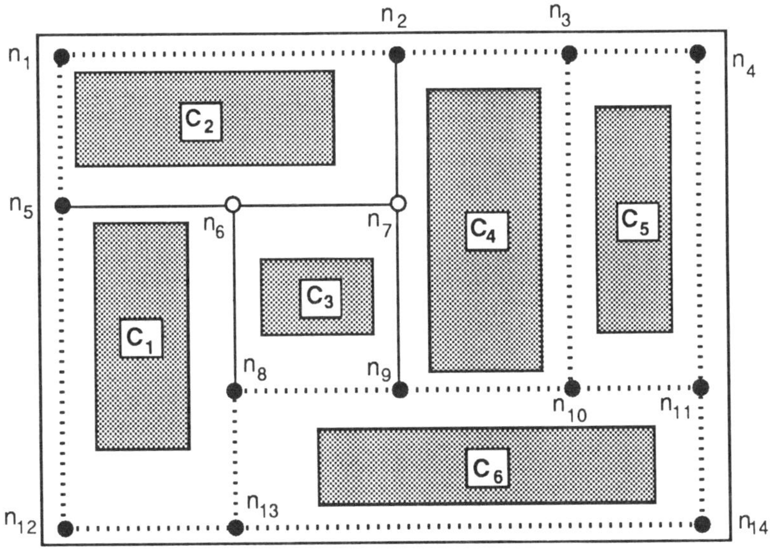

Figure 7.20 The step for $\gamma = 2$.

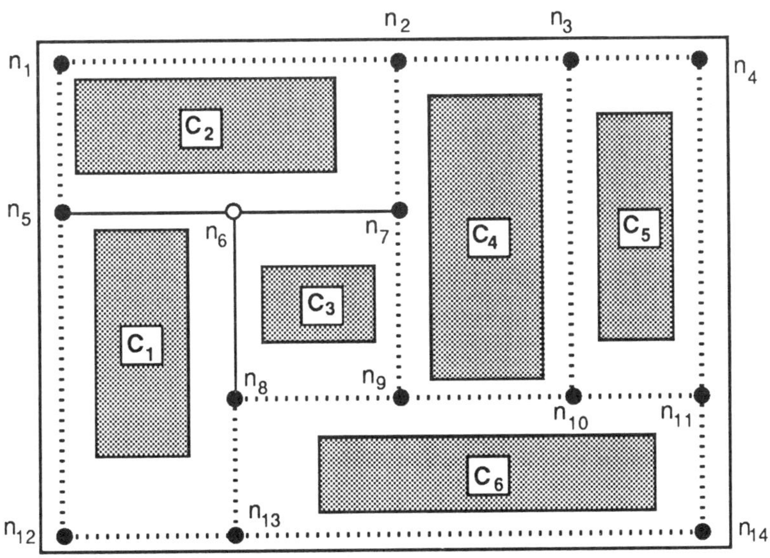

Figure 7.21 The step for $\gamma = 3$.

The step for $\gamma = 4$ is shown in Figure 7.22. The path generated is $P_8 = (n_5, n_6, n_7)$. The set S is shown as dark nodes.

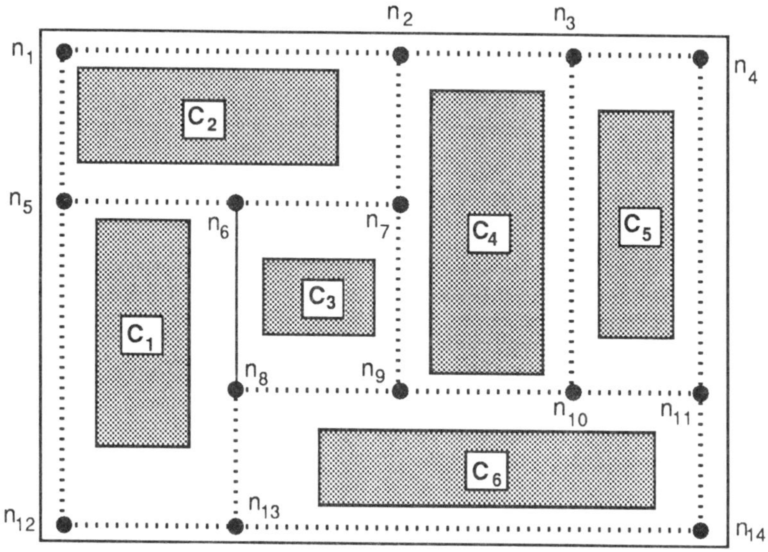

Figure 7.22 The step for $\gamma = 4$.

The step for $\gamma = 5$ is shown in Figure 7.23. The path generated is $P_9 = (n_6, n_8)$.

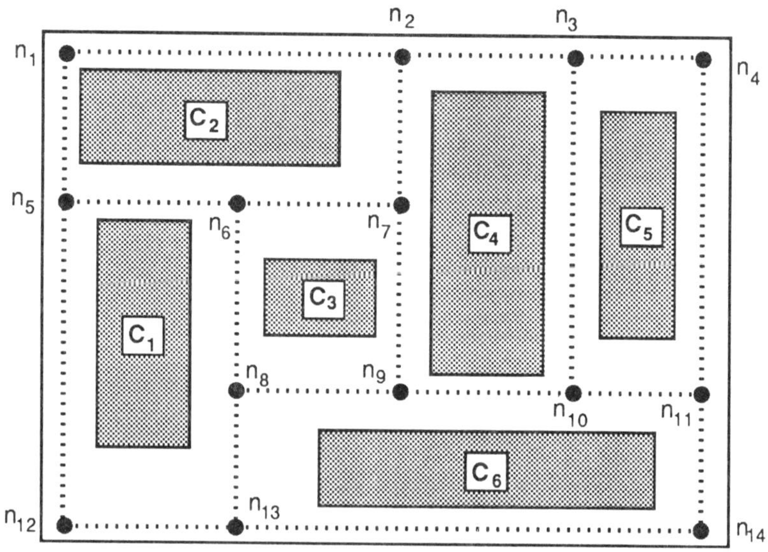

Figure 7.23 The step for $\gamma = 5$.

For the example of Figure 7.23, the paths generated by step 2 above are shown in Table 7.2.

Channel Routing Order
$P_9 = (n_6, n_8)$
$P_8 = (n_5, n_6, n_7)$
$P_7 = (n_2, n_7, n_9)$
$P_6 = (n_3, n_{10})$
$P_5 = (n_{13}, n_8, n_9, n_{10}, n_{11})$
$P_4 = (n_4, n_{11}, n_{14})$
$P_3 = (n_1, n_5, n_{12})$
$P_2 = (n_{12}, n_{13}, n_{14})$
$P_1 = (n_1, n_2, n_3, n_4)$

Table 7.2 Channel routing order for the example of Figure 7.23.

Path P_9 will be handed to the generalized channel router first. When the detailed router finishes, cells C_1 and C_3 will be joined as a unit. Path P_8 is subjected to the detailed router next. Note that the routing of this path joins the previously interconnected cells C_1 and C_3 with cell C_2.

Thirdly, path P_7 is given to the generalized channel router. The routing of this path interconnects cells C_1, C_2 and C_3 with cell C_4. After path P_6 has been routed, cells C_1 through C_5 are joined as a unit. Next, after the routing of path P_5, all of the cells are joined as a unit. At this point, all internal channels have been detail routed. The routing of the external channels, represented by paths P_4, P_3, P_2 and P_1, completes the detailed routing.

For the example of Figure 7.9, the paths generated by step 2 above are shown in Table 7.3.

Channel Routing Order
$P_{11} = (n_{18}, n_{19}, n_{20})$
$P_{10} = (n_5, n_{16}, n_{17})$
$P_9 = (n_7, n_6, n_{14})$
$P_8 = (n_9, n_{10}, n_{13})$
$P_7 = (n_4, n_5)$
$P_6 = (n_8, n_9)$
$P_5 = (n_2, n_5, n_9, n_{12})$
$P_4 = (n_3, n_{20}, n_7, n_{15})$
$P_3 = (n_1, n_4, n_8, n_{11})$
$P_2 = (n_{11}, n_{12}, n_{13}, n_{14}, n_{15})$
$P_1 = (n_1, n_2, n_{17}, n_{18}, n_3)$

Table 7.3 Channel routing order for the example of Figure 7.9.

Paths P_{11} and P_{10} are given first to the generalized channel router. The routing of these paths implements the interconnections along the lower right and lower left portions of cell C_4, respectively. The routing of path P_9 is next. This path joins cell C_5 with the larger cell C_4. The detail routing of path P_8 is next, which interconnects cell C_1 with cell C_4. After routing path P_7, cell C_3 is joined to cell C_2. The routing of path P_6 instantiates the interconnections along the top of cell C_2. Finally, the last internal path is routed, namely path P_5. The routing of this path fully joins the previously interconnected pair C_2 and C_3, with the previously interconnected cells C_1, C_4 and C_5. The routing of the external paths, P_4, P_3, P_2 and P_1, completes the detailed routing for the example of Figure 7.9.

Chapter 8

A Graph-Based Global Router Algorithm

8.1 Introduction

This chapter presents a general-purpose, graph-based global router algorithm. The quality of the solution produced by the global router is independent of the routing order of nets, a common limitation among previous algorithms. The global router is independent of the layout style since the only inputs to the algorithm are a net list and a channel graph (such as that generated by the algorithm of Chapter 7). In the input to the global router, each pin in the net list has been assigned to a specific position on a channel edge in the graph, including electrically equivalent pins. The global router makes full use of equivalent pins to minimize the routing length of a net. The global router minimizes the sum of the routing lengths of all of the nets subject to the satisfaction of the *capacity constraints* of the edges. The constraints result from the fixed widths of the channel edges.

The global router algorithm presented in this chapter has two basic stages. The first stage generates M alternative routes for each net. The parameter M is typically on the order of 20 or more. The algorithm attempts to find the M shortest routes for each net. An efficient algorithm is available for two-pin nets,[1] and I developed an extension which generalizes this approach to handle nets consisting of more than two pins. Results have shown that for nearly all nets of fewer than 20 pins, this algorithm appears to find the minimum Steiner length route included among the M alternatives. The second stage of the global router algorithm selects a single route from the M alternatives for each net. Let n_i represent a net, where $i \in \{1, \cdots, N\}$ and where N is the number of nets. Furthermore, let n_i^k represent the k-th alternative route for net n_i, where $k \in \{1, \cdots, M\}$. A random in-

---

[1] E. Lawler, "Combinatorial Optimization: Networks and Matroids," (1976).

terchange algorithm is then used to select alternative $n_i^{k_0}$ for each $i \in \{1, \cdots, N\}$ while seeking to minimize the total routing length subject to the channel-edge capacity constraints. This approach enables the global router to avoid the net-routing-order dependence problem.

A good review of global router strategies has been published.[1] The basic problem confronting a global router may be stated as follows: Minimize the sum of the routing lengths of all of the nets subject to the satisfaction of the *capacity constraints* of the edges. The constraints result from the fixed widths of the channel edges. The usual approach to global routing is sequential routing. That is, one net at a time is routed. With this approach, the shortest possible route is selected whenever channel-edge capacities permit it.

It is generally not possible to route all nets with the shortest possible interconnection paths because the nets compete with each other for the channel edges to be used. Two of the more common approaches directed toward the solution to this problem are: (1) Route all nets independently, and if some channel edges become excessively congested, reroute the nets causing the excess congestion.[2] (2) Route nets sequentially, and after each net is routed, update the information about channel-edge congestion. Fully utilized channel edges are no longer available for the routing of future nets. Obviously, the routing order becomes a crucial factor. Arguments can be made for routing the shortest nets first, and also for routing the nets with the most pins first. It appears impossible to devise a general procedure which works the best in all situations. Vecchi and Kirkpatrick introduced the use of simulated annealing for global routing in order to avoid a routing-order dependence problem.[3] However, their implementation was also restricted to two-pin nets and only explored L-shaped or Z-shaped routes. Furthermore, their method assumed that all pin locations resided at the intersection of channel edges, that is, at the channel graph nodes.

Most of the previous approaches which avoid the net-routing-order dependence problem restrict the net list to contain only two-pin nets. Also, the shapes of the routes are limited to a few feasible patterns. Usually only one or two bends per net are allowed. With these restrictions, a successful integer programming approach has been reported.[4] Recently, an integer linear programming approach has been applied to the global routing problem.[5] [6] [7] The advantage of this formulation is that the problem instance is viewed in its entirety, and does not depend on any sequential ordering of the nets. An efficient procedure was developed for approximately solving such an integer program. Furthermore, the approximation is provably good in the sense that the solution is guaranteed to be close

---

[1] E. S. Kuh and M. Marek-Sadowska, "Global Routing," (1986).

[2] R. Prim, "Shortest Connecting Networks and Some Generalizations," (1957).

[3] M. Vecchi and S. Kirkpatrick, "Global Wiring by Simulated Annealing," (1983).

[4] R. Karp, R. Leighton, R. Rivest, C. Thompson, U. Vazirani, and V. Vazirani, "Global Wire Routing in Two Dimensional Arrays," (1983).

[5] A. Ng, P. Raghavan, and C. Thompson, "A Specification Language for Describing Rectilinear Steiner Tree Configurations," (1986).

[6] P. Raghavan and C. Thompson, "Provably Good Routing in Graphs: Regular Arrays," (1985).

[7] A. Ng, Raghavan, and C. Thompson, "Experimental Results for a Linear Program Global Router," (1987).

to the best possible. Finally, the procedure runs in polynomial time. A drawback to this approach is that multipin nets are not always handled correctly. For example, the method can produce *track sharing* or *routing cycle* problems for multipin nets.[1]

A newer method, also based on integer linear programming, has been developed with multipin nets in mind.[2] However, while the method appears to work well for three pin nets, it may not be practical for nets with more than three pins. In practice, the two-pins per net restriction and even a three-pins per net limit are very unrealistic, and the search has continued for good heuristics to handle multipin nets, which usually comprise a significant portion of the nets.

As described in Chapter 3, the use of simulated annealing was independently introduced for global routing in the standard cell layout style. This method handles multipin nets and also accounts for precise pin locations. The algorithm utilizes arbitrarily shaped routes and features full utilization of electrically equivalent pins.

The desire to extend these concepts to the case of global routing on a general channel-edge graph provided the motivation for the development of the algorithm presented in this chapter. The algorithm described here has the following features: (1) It is a general-purpose global router which avoids the classical net-routing-order dependence problem. (2) The global router handles nets with any number of pins. (3) The global router fully utilizes electrically equivalent pins to minimize the routing length of a net. (4) The global router is independent of the layout style since the only inputs to the algorithm are a net list and a channel graph. (5) The global router minimizes the sum of the routing lengths of all of the nets subject to the satisfaction of the *capacity constraints* of the edges.

This chapter is organized in the following manner: Section 8.2 reviews some of the basic graph algorithms which are used by the global router. In Section 8.3, the algorithm for generating M-shortest routes for a net is presented. The global router algorithm is then presented in Section 8.4. Finally, in Section 8.5, the results of the global router are compared against manual and automatic techniques for two large industrial circuits.

8.2 Basic Graph Algorithms Used by the Global Router

This section reviews some of the basic graph algorithms used by the global router algorithm described in this chapter. The three key algorithms which form the foundation for the global router algorithm are: (1) Prim's algorithm for finding the minimum spanning tree on a graph,[3] (2) Dijkstra's algorithm for the shortest path on a graph,[4] and (3) Lawler's algorithm for finding the M-shortest paths between two nodes on a graph.[5]

---

[1] A. Ng, Raghavan, and C. Thompson, "Experimental Results for a Linear Program Global Router," (1987).

[2] P. Raghavan and C. Thompson, "Multiterminal Global Routing: A Deterministic Approximation Scheme," (1987).

[3] R. Prim, "Shortest Connecting Networks and Some Generalizations," (1957).

[4] E. Dijkstra, "A Note on Two Problems in Connection with Graphs," (1959).

[5] E. Lawler, "Combinatorial Optimization: Networks and Matroids," (1976).

8.2.1 Prim's algorithm for the minimum spanning tree problem

The input to a minimum spanning tree (MST) algorithm is a graph G, with a weighted set of edges represented by E_G and the set of nodes represented by V_G. Let N represent the cardinality of the set V_G, that is, $|V_G| = N$. Furthermore,

$$V_G = \{ v_g \} = \{ v_g^1, v_g^2, \cdots, v_g^N \}$$

The output is a set of edges defining the MST of G or the message that G is disconnected. Let E_T represent the set of edges of the MST and let $V_T = \{ v_t \}$ represent the set of nodes. Furthermore,

$$V_T = \{ v_t \} = \{ v_t^1, v_t^2, \cdots, v_t^k \}$$

where k is the number of vertices in the MST. The notation e_{ij} will be used to represent an edge connected between nodes i and j. In addition, $w(e_{ij}) \geq 0$ represents the weight of edge e_{ij}. The MST algorithm due to Prim is as follows:

```
MST( G ) {
        E_T = {} ;
        V_T = {} ;
        randomly select: v ∈ V_G ;
        V_T = V_T + v ;
        while( |E_T| < (N − 1) ) {

                e_min = { e_ij |  min over i ∈ {v_t}, j ∈ {v_g} of W( e_ij ) }

                if( e_min == {} ) {
                                /* G is disconnected */
                        return( Tree_Not_Found ) ;
                } else {
                        E_T = E_T + e_min ;
                }
        }
        output( MST ) ;   /* Output the MST of G */
        return( Tree_Found ) ;
}
```

8.2.2 Dijkstra's algorithm for the shortest path problem

One of the inputs to a shortest path (SP) algorithm is a graph G, with a weighted set of edges represented by E_G and the set of nodes represented by V_G. Let N represent the cardinality of the set V_G, that is, $|V_G| = N$. Furthermore,

$$V_G = \{ v_g \} = \{ v_g^1, v_g^2, \cdots, v_g^N \}$$

The other inputs to the algorithm are the source node $s \in V_G$ and the target node $t \in V_G$. The output is a series of nodes, starting from s and ending with t, representing the shortest path from s to t on the graph G.

The notation $l(v_i, v_j)$ represents the length of edge $e_{ij} = (v_i, v_j)$. If (v_i, v_j) is not an edge in E_G and if $i \neq j$, then $l(v_i, v_j) \rightarrow \infty$. Furthermore, $l(v, v) = 0$.

The *SP* algorithm due to Dijkstra has the following basic strategy: Construct a set $S \subseteq V_G$ such that the shortest path from s to each vertex $v \in S$ lies wholly in S. The array $D[v]$ contains the cost of the current shortest path from s to v passing only through vertices of S. The algorithm is as follows:

```
SP( G , s , t ) {
        S = { s } ;
        D[s] = 0 ;
        for( each v ∈ { V_G − S } ) {
                D[v] = l( s , v ) ;
        }
        while( S ≠ V_G ) {
                w =     min       D[v]
                    v ∈ {V_G−S}
                S = S + w ;
                for( each v ∈ { V_G − S } ) {
                        D[v] = min { D[v] , D[w] + l( w , v ) } ;
                }
        }
}
```

8.2.3 Lawler's algorithm for finding the M-shortest paths

As implied in Section 8.1, I desire to compute not only the shortest path, but the $(M-1)$ ***next shortest*** paths between a specified pair of nodes in a graph G. The algorithm for finding the M-shortest paths in a graph, that is, the *MSP* algorithm, is due to Lawler.[1] Each of the M paths comprising the output of the *MSP* algorithm is not permitted to have repeated nodes. As opposed to Section 8.2.2, the source node will be referred to as node 1 and, without loss of generality, the target node will be represented by node n, where n is the number of nodes in G. The length of each edge in the graph is assumed to be nonnegative.

The structure of the algorithm is given below. Note that the *MSP* algorithm uses the *SP* algorithm of the previous subsection. The algorithm uses a priority queue Q to store the list of shortest paths from node 1 to n which have been found up to any point in the execution of the algorithm. A shortest path is characterized by its length. If at any point in the algorithm Q contains $k > M$ paths, the $(k - M)$ longest paths are purged. Without loss of generality, assume that P_m contains edges (1,2), (2,3), $\cdots$, $(q\text{-}1, q)$, (q, n) and that P_m is the shortest path from node 1 to node n subject to the conditions that it is forced to in-

[1] E. Lawler, "Combinatorial Optimization: Networks and Matroids," (1976).

clude edges (1,2), (2,3), $\cdots$, $(p\text{-}1, p)$, where $p \le q$, and that certain edges from node p are excluded. These conditions are stored with P_m as part of the entry in Q. Note that node n is the same as node $q + 1$.

The call to the function SP is represented by the notation: $SP_p^k(G, 1, n, X_p)$. This notation implies that Dijkstra's method is to find the shortest path from node 1 to node n on the graph G subject to the conditions that edges (1,2), (2,3), $\cdots$, $(p\text{-}1,p)$ are included. A second condition implied by the notation is that edge $(k, k+1)$ is to be excluded. The list X_p, if not empty, includes previously excluded edges from node p for path P_m (the previously generated path) which must also be excluded. The function $SP_p^k(G, 1, n, X_p)$ returns null if no such path was found.

```
MSP( G , 1 , n , M ) {
        PUSH( Q ) = SP( G , 1 , n ) ;
        for( m = 1; ; m = m + 1 ) ;
                if( Q == {} ){
                        return ;            /* there are no more paths */
                } else {
                        P_m = POP( Q );
                }
                if( m == M ){
                        return ;            /* the computation is complete */
                }
                PUSH( Q ) = SP^p_p( G , 1 , n , X_p )
                        /* If an entry was added to Q, a record of the conditions under which it was ob-
                        tained is stored as part of the entry */
                if( p < q ){
                        for( c = 1 ; c < q - p ; c++ ){
                                PUSH( Q ) = SP^{p+c}_{p+c}( G , 1 , n , { } ) ;
                        }
                }
        }
}
```

The time complexity of Dijkstra's SP algorithm is $O(n^2)$. In order to compute the M-th shortest path, $O(Mn)$ shortest path computations, each of $O(n^2)$ must be carried out. Hence, $O(Mn^3)$ running time is required for these computations. The algorithm may generate as many as $O(Mn)$ entries for Q. Since the value of M is known in advance, all but the shortest $(M\text{-}m)$ paths in Q can be discarded at the m-th iteration. Thus the number of entries in Q is $O(M)$ and since each entry requires $O(n)$ space, the total storage requirement for Q is $O(Mn)$.

8.3 The Algorithm for Generating M-Shortest Routes for a Net

This section presents the extension of Lawler's algorithm to the case of multipin nets on a graph, including electrically equivalent pins. The algorithm is referred to as the *MSR* (*M*-shortest routes) algorithm. This section is organized in the following way: First, the *MSR* algorithm is presented. Then, an example will be shown to demonstrate the operation of the algorithm.

```
MSR() {
    for( net = 1 ; net ≤ N ; net++ ) {   /* N is the number of nets */
        G = G + pins[net] ;
            /* The array pins contains the pins for net. Add the pins and electrically
            equivalent pins (as nodes) for net to the channel-edge graph G. */
        i = random_integer( pins[net] ) ;
            /* i is a randomly chosen node of net */
        list_of_source_nodes = { 1 , ··· , pins[net] } ;
        list_of_target_nodes = {} ;
        recurse_path( list_of_source_nodes, { i } , list_of_target_nodes, { i } ) ;
        G = G – pins[net] ;
            /* Examination of net is complete. Remove the pins (nodes) from graph G. */
    }
}
```

```
recurse_path( list_of_source_nodes , delete_from_source_list ,
                                        list_of_target_nodes , add_to_target_list ) {
    if( list_of_source_nodes - delete_from_source_list == {} ) {
        save_path() ;
            /* The function save_path() stores the path defined by list_of_target_nodes +
            add_to_target_list in a priority queue. The figure of merit used by the priority
            queue is the length of the path. If the M-shortest paths are desired, the queue is
            purged of the longest paths whenever it contains more than M paths. */
        return ;
    }
    list_of_source_nodes = list_of_source_nodes - delete_from_source_list ;
    list_of_target_nodes = list_of_target_nodes + add_to_target_list ;
    for( each i ∈ list_of_source_nodes ) {
        PUSH( sQ ) = { i | min_{j ∈ list_of_target_nodes} distance( i , j ) }
            /* The priority queue sQ stores the node set list_of_source_nodes in increasing
            order of distance to the current target node set list_of_target_nodes. In practice,
            the number of source nodes considered is usually limited by a parameter which is
            used to control the size of the search space. */
    }
    while( sQ ≠ {} ) {
        i = POP( sQ ) ;
        for( each j ∈ list_of_target_nodes ) {
                /* In practice, the number of target nodes to be considered is limited to
                the M nearest nodes. Those target nodes which are farther from node i
                than the M nearest target nodes could not possibly give rise to one of
                the M-shortest paths. */
            PUSH( Q ) = MSP( G , i , j , M ) ; /* Q is a priority queue */
        }
    }
    for( c = 1 ; c ≤ M ; c++ ) {
        Paths[c] = POP( Q ) ;
    }
        /* For any path p ∈ Paths[c] , c = 1, ··· , M, the function Nodes( p ) returns the list of
        nodes defining the path p. The function source_nodes( Nodes( Paths[c] ) ) returns the list
        of additional source nodes included by the path p.*/
    for( c = 1 ; c ≤ M ; c++ ) {
        recurse_path( list_of_source_nodes , source_nodes( Nodes( Paths[c] ) ) ,
                                list_of_target_nodes , Nodes( Paths[c] ) ) ;
    }
    list_of_source_nodes = list_of_source_nodes + delete_from_source_list ;
    list_of_target_nodes = list_of_target_nodes - add_to_target_list ;
    return ;
}
```

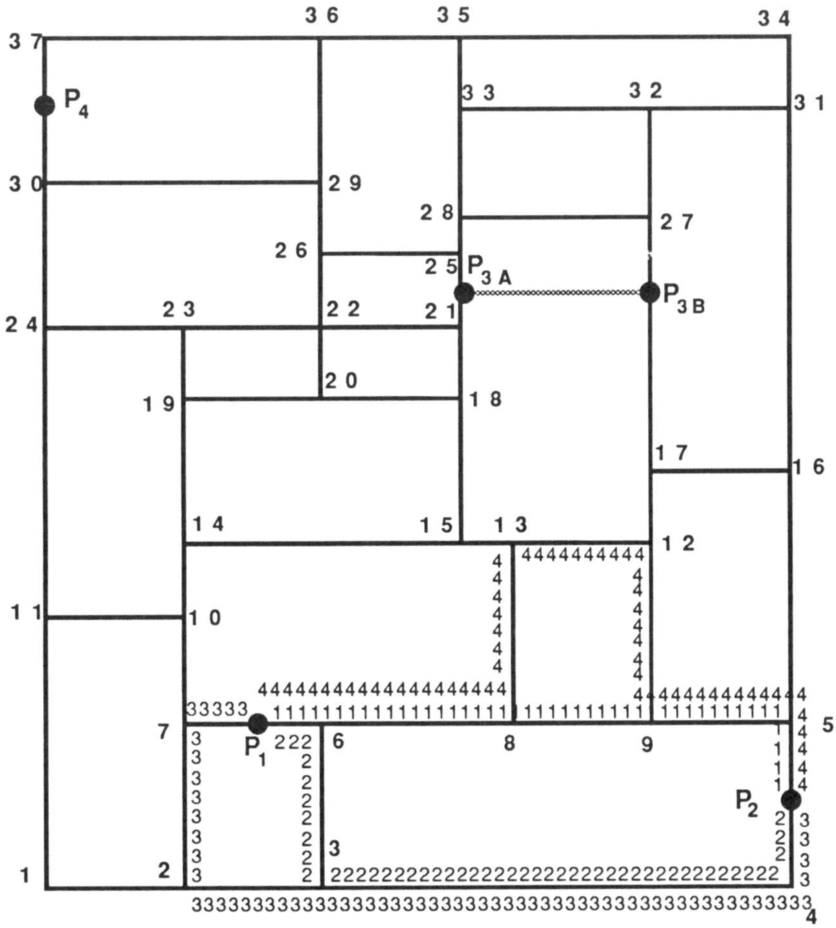

Figure 8.1 A five-pin net is to be routed on the given graph. There are only four distinct pin groups, as pins P_{3A} and P_{3B} are electrically equivalent. Four of the paths for the outermost level of recursion are shown. The path number (ranked inversely according to length) is repeated along the edges in the channel-edge graph to indicate the path. Path 1 (stored in *Paths[1]*) proceeds from node P_1 to nodes 6, 8, 9, 5 and P_2. Path 2 (stored in *Paths[2]*) takes the following route: node P_1 to nodes 6, 3, 4 and P_2. Similarly, path 4 (stored in *Paths[4]*) proceeds from node P_1 to nodes 6, 8, 13, 12, 9, 5 and P_2.

The example in Figure 8.1 is that of a five-pin net. However, there are only four distinct pin groups, as pins P_{3A} and P_{3B} are electrically equivalent. It is assumed that pin $i = P_2$ is the first pin selected by the *MSR* algorithm. The first call to the function recurse_path() is made with the following arguments:

list_of_source_nodes = { P_1 , P_2 , P_{3A} , P_{3B} , P_4 }
delete_from_source_list = { P_2 }
list_of_target_nodes = { }
add_to_target_list = { P_2 }

Node P_2 is then deleted from the *list_of_source_nodes* and added to the *list_of_target_nodes*. The call to the *min* function in recurse_path() then yields $i = P_1$ as the closest source node. (For this part of the example, only the closest source node will be considered, although in general others would be considered as well.) The *MSP* function is then called to generate the *M*-shortest paths between nodes P_1 and P_2. In Figure 8.1, four of the paths are shown. The path number (ranked inversely according to length) is repeated along the edges in the channel-edge graph to indicate the path. Path 1 (stored in *Paths[1]*) proceeds from node P_1 to nodes 6, 8, 9, 5 and P_2. Path 2 (stored in *Paths[2]*) takes the following route: node P_1 to nodes 6, 3, 4 and P_2. Similarly, path 4 (stored in *Paths[4]*) proceeds from node P_1 to nodes 6, 8, 13, 12, 9, 5 and P_2. The algorithm selects, in turn, each of the *M* paths stored in the array *Paths* and recursively calls the function recurse_path(). The fourth such recursive call to recurse_path() will have the following arguments:

list_of_source_nodes = { P_1 , P_{3A} , P_{3B} , P_4 }
delete_from_source_list = { P_1 }
list_of_target_nodes = { P_2 }
add_to_target_list = { P_1 , *6* , *8* , *13* , *12* , *9* , *5* }

First, node P_1 is deleted from *list_of_source_nodes*. Second, nodes P_1 , 6, 8, 13, 12, 9 and 5 are added to *list_of_target_nodes*. The call to the *min* function in recurse_path() yields P_{3A} and P_{3B} as the closest source-node pair. The *MSP* function is then called to generate the *M* shortest paths between nodes P_{3A} and the nodes in *list_of_target_nodes* as well as between nodes P_{3B} and the nodes in *list_of_target_nodes*.

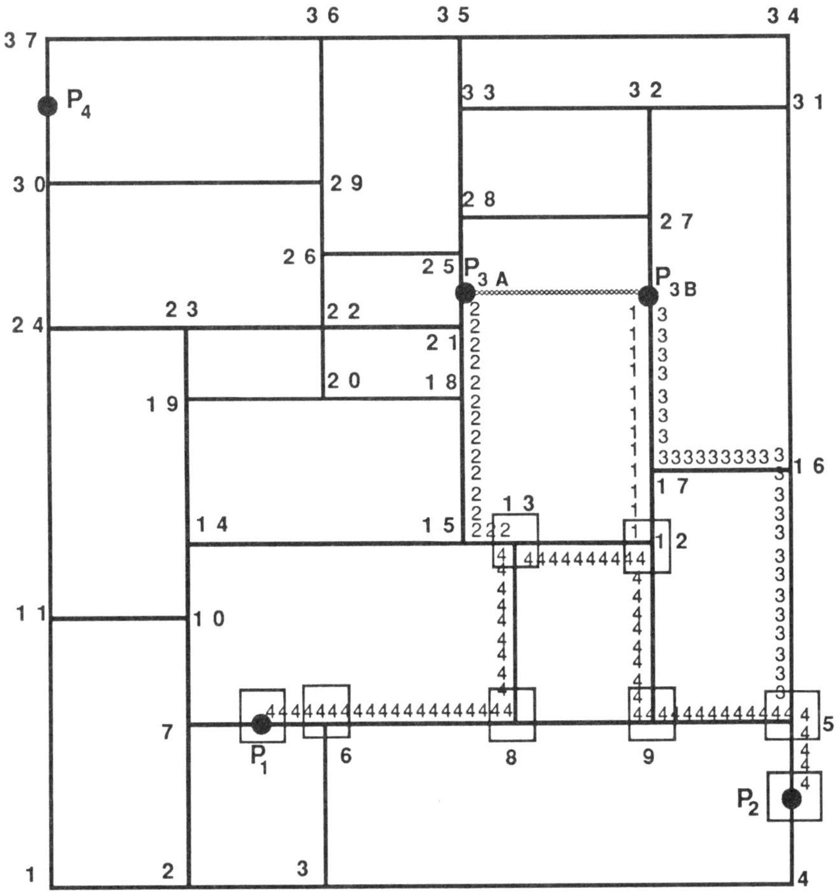

Figure 8.2 Three paths are shown for the second level of recursion. The path number (ranked inversely according to length) is repeated along the edges in the channel-edge graph to indicate the path. Path 1 (stored in *Paths[1]*) proceeds from node P_{3B} to nodes 17 and 12. Path 2 (stored in *Paths[2]*) takes the following route: node P_{3A} to nodes 21, 18, 15 and 13. Similarly, path 3 proceeds from node P_{3B} to nodes 17, 16, and 5. Each node in *list_of_target_nodes* is encompassed by a small rectangle.

In Figure 8.2, three of the paths are shown. The path number (ranked inversely according to length) is repeated along the edges in the channel-edge graph to indicate the path. Each node in *list_of_target_nodes* is encompassed by a small rectangle. Path 1 (stored in *Paths[1]*) proceeds from node P_{3B} to nodes 17 and 12. Path 2 (stored in *Paths[2]*) takes the following route: node P_{3A} to nodes 21, 18, 15 and 13. Similarly, path 3 (stored in *Paths[3]*) proceeds from node P_{3B} to nodes 17, 16 and 5.

The algorithm selects, in turn, each of the M paths stored in the array *Paths[]* and recursively calls the function recurse_path(). The second such recursive call to recurse_path() will have the following arguments:

list_of_source_nodes = { P_{3A} , P_{3B} , P_4 }
delete_from_source_list = { P_{3A} , P_{3B} }
list_of_target_nodes = { P_1 , *6* , *8* , *13* , *12* , *9* , *5* , P_2 }
add_to_target_list = { P_{3A} , P_{3B} , *21* , *18* , *15* }

First, nodes P_{3A} and P_{3B} are deleted from *list_of_source_nodes*. Second, nodes P_{3A}, P_{3B} , 21, 18, and 15 are added to *list_of_target_nodes*. Since there is only one source node remaining in *list_of_source_nodes*, the call to the *min* function in recurse_path() is done only for source node P_4. The *MSP* function is then called to generate the M shortest paths between nodes P_4 and the nodes in *list_of_target_nodes*. In Figure 8.3, three of the paths are shown.

The path number is repeated along the edges in the channel-edge graph to indicate the path. Path 1 (stored in *Paths[1]*) proceeds from node P_4 to nodes 30, 29, 26 and 25. Path 2 takes the following route: node P_4 to nodes 30, 29, 26, 22 and 21. Similarly, path 5 proceeds from node P_4 to nodes 37, 36, 35, 33, 32 and 27.

Each time *list_of_source_nodes* is emptied, the function save_path() is called to store the path defined by: *list_of_target_nodes* + *add_to_target_list* in a priority queue. The figure of merit used by the priority queue is the length of the path. Since the M-shortest paths are desired, the queue is purged of the longest paths whenever it contains more than M paths. The recursion of the calls to recurse_path() will come to a close following the examination of all of the M paths which were stored between the first pair of nodes, namely P_1 and P_2.

For net n, the variable G_n is defined to be the number of electrically independent pin groups. The net illustrated in Figures. 8.1 through 8.3 has $G_n = 4$ since pins P_{3A} and P_{3B} are electrically equivalent. The number of recursive calls to recurse_path() can then be expressed as M^{G_n}. That is, the *MSR* algorithm is to select the M shortest of the M^{G_n} routes that are generated.

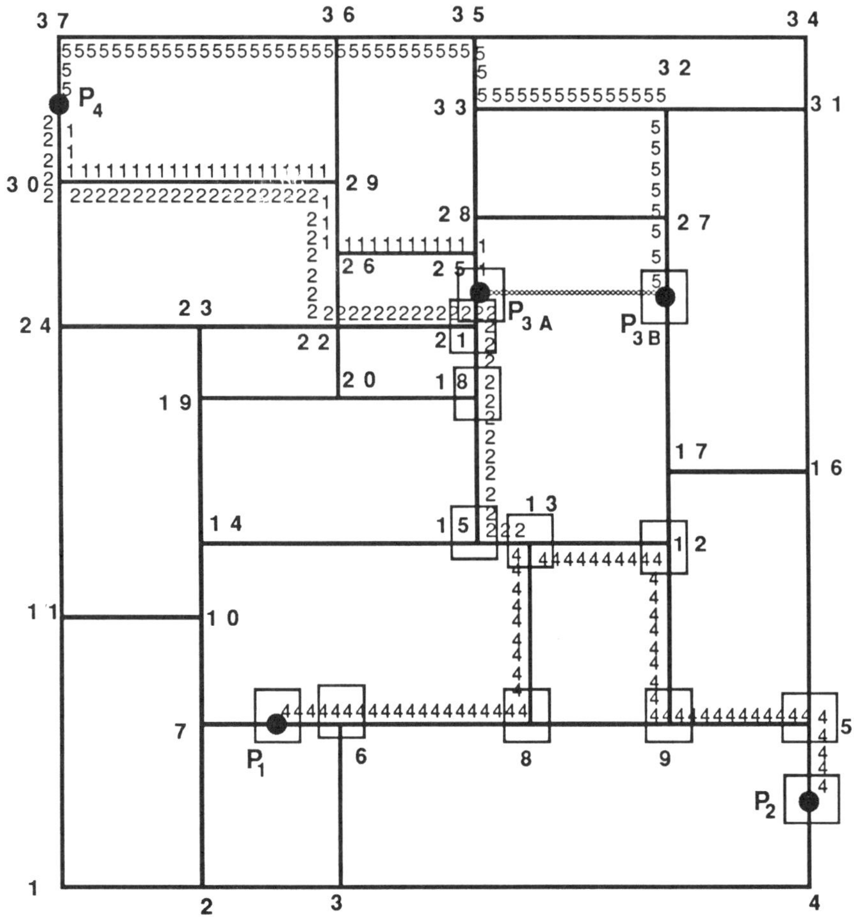

Figure 8.3 Three paths are shown for the third level of recursion. The path number (ranked inversely according to length) is repeated along the edges in the channel-edge graph to indicate the path. Path 1 (stored in *Paths[1]*) proceeds from node P_4 to nodes 30, 29, 26 and 25. Path 2 (stored in *Paths[2]*) takes the following route: node P_4 to nodes 30, 29, 26, 22 and 21. Similarly, path 5 (stored in *Paths[5]*) proceeds from node P_4 to nodes 37, 36, 35, 33, 32 and 27. Each node in *list_of_target_nodes* is encompassed by a small rectangle.

Each call to recurse_path() involves up to M calls to the MSP algorithm, which is $O(MN^3)$, where N is the total number of nodes, including the nodes generated as a result of the pins for a given net. Thus, each call to recurse_path() has time complexity $O(M^2 N^3)$. For M^{G_n} calls to recurse_path(), the overall time complexity is given by: $O(M^{G_n} M^2 N^3)$. Simplification yields the following for the time complexity:

$O\left(M^{G_n+2}N^3\right)$

For $M \approx 20$, the amount of calculation necessary can be excessive for $G_n \geq 6$. That is, for larger nets ($G_n \geq 6$), the number of recursive calls to recurse_path() must be limited. This is accomplished by selecting $M_0 < M$ for these nets such that the number of recursive calls expressed by M^{G_n} is manageable. Since $M \leq 30$ is usual, and since $M_0 < M$ is usually selected such that $M^{G_n} \gg M$, we should not expect to observe a reduction in quality of the M-shortest routes that are found.

8.4 The Second Phase of the Global Router Algorithm

The second phase of the algorithm selects a single route from among the $M_i \leq M$ alternatives for each net i, where $i \in \{1, 2, \cdots, N\}$ and N is the number of nets. Note that M_i is specified as $M_i \leq M$ since cases arise in which fewer than M alternative routes are possible for a given net. Let i_k represent the k-th alternative route for net i, where $k \in \{1, 2, \cdots, M_i\}$. A random interchange algorithm is then used to select alternative i'_k for each net while seeking to minimize the total routing length subject to the channel-edge capacity constraints. This approach enables the global router to avoid the net-routing-order dependence problem.

This section describes the cost and generate functions for the random interchange algorithm. The cost function is represented by:

$$L = \sum_{i=1}^{N} \lambda(i_k) \tag{8.1}$$

where $\lambda(i_k)$ represents the length of alternative route i_k.

Let E_j represent a channel edge, where $j \in \{1, 2, \cdots, N_{ch}\}$ and where N_{ch} represents the total number of channel edges for a circuit. The density of channel edge j is represented by D_j and the capacity of the edge is represented by C_j. Given that X represents the total number of excess tracks required over all channel edges, a second cost function is defined by:

$$X = \sum_{j=1}^{N_{ch}} \left\{ (D_j - C_j) \mid D_j > C_j \right\} \tag{8.2}$$

The M_i possible routes for each net i are enumerated such that the $k = 1$ route is the shortest. That is, i_1 is the shortest route for each net i. The first state of the random interchange algorithm is such that $k = 1$ for each net i. If this first state is such that no capacity-constraint violations are observed, that is, if there does not exist $j \in \{1, 2, \cdots, N_{ch}\}$ such that $D_j > C_j$, then the algorithm terminates since each net has been given its shortest route with no capacity violations. If capacity-constraint violations are observed, the random interchange algorithm visits other states while attempting to eliminate the excess congestion.

New state generation begins with the random selection of a channel edge E_j characterized by $D_j > C_j$. A net i containing a segment using E_j is then randomly selected. An alternative route for net i is selected randomly from among those characterized by $\Delta X \leq 0$. If $\Delta X = 0$ and $\Delta L \leq 0$, or if $\Delta X < 0$, then the new route is accepted; otherwise it is rejected.

The *stopping criterion* is satisfied when either: (1) each net has been assigned the k=1 route and X=0, or (2) the values of L and X have not changed for $M{\cdot}N$ new state attempts.

8.5 Results

In this section the performance of the general global router is compared with automatic and manual methods. The measure for comparison was only total interconnect length, hence the random interchange portion of the algorithm, used to eliminate excess congestion, was not used. That is, it was assumed in the comparative tests that the capacity of each edge in the channel-edge graph was not bounded.

As was mentioned in Section 8.3, practical considerations with respect to run time force the selection of $M_0 < M$ for $G_n \gg 1$. In the current implementation of the global router, the following expression is used to compute M_0:

$$M_0 = \begin{cases} \left\lfloor \dfrac{M}{(G_n - 1)} \right\rfloor & \text{if } G_n > 2 \\ M & \text{if } G_n \leq 2 \end{cases} \tag{8.3}$$

Equation 8.3 indicates that for M = 20, the number of alternative routes generated between each pin group in the set of G_n pin groups is exactly one (M_0 = 1) for $G_n \geq 12$. Other formulations were not tested for large nets since relatively few nets are characterized by $G_n \geq 12$. Note that for M = 20 in conjunction with $G_n \geq 12$, the operation of the generalized global router essentially follows a generalization of Prim's algorithm for the MST of the net. The generalization being that potential Steiner (target) points are added whenever a new pin is added to nodes of the spanning tree. Therefore it would not be expected that the global router would always find the minimum length route for nets characterized by $G_n \geq$ 12.

Characteristics of the two large circuits which were used for comparison to automatic and manual methods are shown in Table 8.1. No electrically equivalent pins were present in the net list of either circuit, implying that a key aspect of the capability of the generalized global router was not required.

Circuit	No. Cells	No. Nets	No. Pins
11	62	570	4309
d2	20	656	1776

Table 8.1 Characteristics of two industrial circuits used to test the generalized global router.

For the tests of the generalized global router, the values of M which were considered were 5, 10, and 20. As mentioned in the description of the recurse_path() algorithm in Section 8.3, the number of source nodes considered at each stage of the *look-ahead* process in the *MSR* algorithm is controlled by a parameter. This parameter was set to 3 for all of the tests. Table 8.2 shows the generalized global router results for circuit d2.

M	CPU time (minutes) (VAX 8650)	Total Interconnect Length
20	95	4,844,980
10	8.5	4,854,286
5	1.0	4,862,266

Table 8.2 The performance of the generalized global router on circuit d2.

It is apparent the run time of the *MSR* algorithm is strongly dependent on *M*. The generalized global router was compared with the NLAGR global router developed recently at UC Berkeley.[1] The NLAGR router is only capable of generating a single route for a given net and does not utilize electrically equivalent pins. Hence a true comparison of the two methods is not possible. However, it is useful to sum the lengths of the shortest of the M_n routes generating for each net n by the generalized global router and compare it with the sum of the lengths of the routes found by NLAGR. The NLAGR result for circuit d2 is shown in Table 8.3.

CPU time (minutes) (VAX 8650)	Manual Effort (man-hours)	Total Interconnect Length
1.2	0	4,853,670
1.2	30	4,844,980

Table 8.3 The performance of the NLAGR algorithm on circuit d2, without and with use of the manual interactive facilities.

The NLAGR router has the capability of permitting full-manual intervention of the global routing process. That is, the global route of a net generated NLAGR can be altered by the user. Hence, it was possible to obtain a total interconnect length result for manual global routing. With the interactive facilities supported by NLAGR, it was possible to globally route each net in an average of 3 minutes, for a total of 30 man-hours to complete the job. The manual result was a total interconnect length value of 4,844,980.[2] Note that the generalized global router yielded the same result in only 95 minutes of computation time. During the 95 minutes, the generalized global router also produced 19 of the next shortest routes for each net.

Table 8.4 shows the generalized global router results for circuit l1. This circuit had 62 cells, in comparison to 20 for circuit d2. Hence, the channel-edge graph was signifi-

---

[1] G. Srinath, "NLAGR," (1986).

[2] G. Srinath, "NLAGR," (1986).

cantly more complex, leading to the longer computation times. Also, circuit 11 had almost three times as many pins per net, as can be seen from Table 8.1.

The NLAGR result for circuit 11 is shown in Table 8.5. With the interactive facilities supported by NLAGR, it was possible to globally route each net in an average of 5 minutes, for a total of 50 man-hours to complete the job. The manual result was a total interconnect length value of 4,090,948. For a total of 31 of the 570 nets, the manual result was better than the result produced by the generalized global router. The breakdown of these nets is shown in Table 8.6.

M	CPU time (minutes) (VAX 8650)	Total Interconnect Length
20	600	4,101,142
10	25	4,113,975
5	5	4,126,055

Table 8.4 The performance of the generalized global router on circuit 11.

CPU time (minutes) (VAX 8650)	Manual Effort (man-hours)	Total Interconnect Length
14.5	0	4,104,246
14.5	50	4,090,948

Table 8.5 The performance of the NLAGR algorithm on circuit 11, without and with use of the manual interactive facilities.

The total difference in interconnect length was 10,194, or less than 0.96 percent. Note that the generalized global router achieved the route found manually for *all* nets with less than 8 pins. The computation time was five times less than the manual effort, and in addition, the general global router produced 19 of the next shortest routes for each of the nets. Consequently, comparing the run time directly with NLAGR is not a fair comparison since NLAGR is only able to generate one possible route per net. This can be a serious problem if channel capacities are limited. The key goal of the first phase of the global router is to generate M of the shortest possible routes for each net, not only to find the one shortest route.

Note from Eqn. 8.3 that the value of M_0 was 1 for the 26 nets represented by the first two rows of Table 8.6. This implies that for these nets, the number of alternative routes generated is also exactly one. In this case, the operation of the generalized global router essentially follows a generalization of Prim's algorithm for the MST of the net. Therefore it would not be expected that the global router would always find the minimum length route for these nets.

Number of Nets	Number of Pins
6	> 600
20	≥ 12
2	10
3	8

Table 8.6 Statistics on the 31 of 570 nets in which manual global routing found a better solution.

From Eqn. 8.3, the value of M_0 was 2 for the nets represented by the third and fourth rows of Table 8.6. Certainly the probability of finding the minimal length route increases as M_0 increases. Thus, if the generalized global router fails to find the minimum length route, it is not surprising that all such occurrences are characterized by $M_0 \leq 2$. Note that for circuit d2, which had substantially fewer pins per net on the average than circuit l1, the generalized global router matched the manual routing length for *each* net.

As presented in Section 8.3, the total number of routes generated for a net is given by $M_0^{G_n+2}$, where only the $M \geq M_0$ shortest are to be retained. Nets with large numbers of pins (large G_n) strongly impact the computation time under the current implementation. It is my opinion that the computation time could be reduced dramatically by slightly changing the manner in which alternative routes are explored. Currently, M_0 routes are generated between every remaining source node and every current target node. During the earlier stages of considering alternatives for a net, when the number of accumulated target nodes is small, it is important to consider many possibilities. However, it is my opinion that as more and more pins become target nodes, then fewer possible connections from the source nodes to the target nodes need to be considered. The quality of the M shortest routes should not be affected since the number of routes explored for each net would still be dramatically greater than M. Also, the current implementation has the drawback that for very large nets, M_0 must be set to one since even 2^{G_n+2} is too large. Thus very few alternatives are explored for the extremely large nets, thereby explaining the performance problems for these nets. It is my opinion that better performance would be obtained by starting with a value of M_0 larger than one, and then gradually decreasing its value toward 1 as most of the pins are included in the target list.

Chapter 9

Conclusion

9.1 Summary

The use of the simulated annealing algorithm enabled the development of automatic placement and global routing algorithms which performed significantly better than a wide variety of other automatic and manual methods. Also, the algorithms based on simulated annealing captured many of the basic requirements and flexibility demanded by industrial layout problems.

The two main characteristics of a simulated annealing algorithm are: (1) a cost function reflecting the quality of the current state, and (2) a new-state generation function allowing the exploration of the configuration space. Such simplicity permits the addition of a great number of problem-specific constraints into the cost function and the generation of new states function. For example, for the automatic placement problem, nonrectangular cells, uniform-height cells with widely varying widths, critical-net weighting, cell exchange classes, and exact pin locations were all accommodated with ease. Furthermore, with respect to global routing, it was possible to consider exact net embeddings and also to eliminate the routing-order dependence problem.

Even while incorporating tremendous flexibility, the simulated annealing algorithm achieves excellent optimization results due to the controlled acceptance of uphill moves. The one drawback to the technique is that a new state is generated in a random, incremental fashion from the current state. The acceptance or rejection of the new state is governed only by the change in the value of the objective function and the value of the temperature parameter. Since this is an undirected search, it is not surprising that optimal performance of the algorithm can only be achieved through the exploration of many more states in the configuration space than would be required by directed-search algorithms. The main focus of this work was to demonstrate that it was possible to devise placement and global routing algorithms which capture all of the flexibility required by state-of-the-art industrial circuits. Furthermore, the performance of these algorithms must match or exceed the performance obtained by directed-search algorithms. Very little attention was paid to minimizing the computation time. The only requirement was that the computation time must be bounded by a polynomial in the size of the problem.

A standard cell placement and global routing package, named TimberWolfSC, was developed based on the simulated annealing algorithm. The package is also applicable to the row-based gate array placement and global routing problem. The cells are permitted to have any shape or size. Macro blocks are also permitted on the chip, and pad placement along the periphery of the core region is also performed. The first stage of the placement process seeks to minimize the total interconnect length. TimberWolfSC then enters a placement-refinement stage, in which local changes are made to the placement should such a change lead to a reduction in the number of wiring tracks required. The computation of the total number of wiring tracks requires a global routing algorithm. An algorithm was developed, based on the total channel density model for row-based layout images. All pins, including electrically equivalent pins, are considered during the global routing process. At any step in the optimization, a feasible global routing for the entire chip is maintained. The key aspect of the new global router is the optimization of the placement of switchable net segments.

Results on industrial circuits versus numerous automatic and manual layout methods showed that TimberWolfSC yielded area savings ranging from 15 to 75 percent. For all circuits tested, the global router reduced the number of wiring tracks needed by an additional 7 to 16 percent in comparison to total interconnect length minimization alone. Furthermore, the placement-refinement stage produced wiring-track reductions ranging from 15 to 25 percent in comparison to total interconnect length minimization alone. TimberWolfSC also features critical-net weighting for performance-driven placement.

A package named TimberWolfMC, also based on simulated annealing, was developed for the macro/custom cell placement and global routing problem. TimberWolfMC is applicable to circuits containing cells of any rectilinear shape. Furthermore, the cells may have fixed geometry including pin locations (macro cells) or the cells may have an estimated area with a specified aspect-ratio range, and with pins that need to be placed (custom cells). All eight possible orientations are considered for each cell. TimberWolfMC also permits the custom cells to have aspect ratios in a continuous or discrete range. Furthermore, several alternative instances for a custom cell may be specified, whereby TimberWolfMC is to select the one which is most suitable. The selection of aspect ratio and/or instance is guided by the minimization of the total interconnect length and by the geometry of the empty space allotted for the cell as influenced by the neighboring cells. TimberWolfMC places circuits consisting entirely of macro cells as well as circuits consisting entirely of custom cells. Furthermore, the program will place circuits consisting of a combination of macro and custom cells. Consequently, TimberWolfMC is applicable to chip planning problems.

The placement algorithm proceeds in two distinct stages. During the first stage, the interconnect area around the individual cells is determined using the dynamic interconnect-area estimator. That is, as each cell is moved, its effective area is adjusted. A simulated annealing algorithm is used to minimize the TEIC. The second stage of TimberWolfMC consists of several executions of a placement-refinement algorithm. Each execution consists of three steps: (1) a channel definition step, (2) a global routing step, and (3) a placement-refinement step. The information obtained in step 2 is used to compute the density of all of the channels, which then permits accurate interconnect-area determination. The placement of the cells is then refined in step 3 to reflect the required interconnect area. A low-temperature simulated annealing algorithm is used.

Total interconnect length savings of 8 to 49 percent were achieved in experiments on 9 industrial circuits. Furthermore, for the 9 examples, circuit-area reductions ranged from 4 to 56 percent versus a variety of other placement algorithms.

An accurate model for estimating the average interconnection length for both random and optimized placements was developed. The model for random placements yielded average interconnection lengths which were typically within 5 percent of the measured values. Furthermore, the model developed for optimized placements enabled the computation of values of ρ, the ratio of the average interconnection length for random over optimized placements, which were typically within 7 percent of the measured ratios. The predicted results were obtained strictly from an analysis of the net list, with no *a priori* knowledge of the functionality of the circuit.

A dynamic interconnect-area estimation algorithm was also developed. The algorithm estimates the interconnect area needed for each edge of a cell. The estimate is based on three factors: (1) the average net traffic, which is an estimate of the average number of interconnections passing through the cross-section of a channel, (2) the relative position of the edge on the chip, and (3) the relative pin density of the edge. Factor (2) is updated each time a cell is moved during the simulated annealing process. The CPU time required to update this factor is minimal. Factors (1) and (3) are determined at the outset and stored.

The effectiveness of the estimator was demonstrated by comparing the final chip area as achieved by the simulated annealing algorithm using the estimator versus the final chip area following a global routing step. Also, the total interconnect length was compared for the two cases, since a disparity would have indicated substantial cell movement was necessary to accommodate the global routing. Test results were reported for 9 industrial macro/custom cell circuits. The final chip areas and the final total interconnect lengths were within a few percent of each other for all 9 examples. The dynamic estimator has enabled the generation of placements which require very little placement modification during or after detailed routing.

A channel definition algorithm for the rectilinear macro/custom cell environment was also developed. This algorithm may be used either prior to detailed routing or prior to the placement-refinement phase. The channel generation procedure differs from conventional channel definition algorithms in that each channel is bordered by at most two cell edges. Hence the new procedure is suitable for placement refinement.

A guaranteed noncyclic routing order is produced for the set of channels with one restriction.[1] This restriction assumes that a generalized channel router is available, that is, a router which can handle channels of $n \geq 4$ sides, with fixed pins on any of the n sides.

[1] A. Sangiovanni-Vincentelli, Private Communication, (1986).

A graph-based global routing algorithm was also developed. This algorithm, used in TimberWolfMC, has the following features: (1) It is a general-purpose graph-based global router which avoids the classical routing-order dependence problem. (2) Multipin nets are handled in the same fashion as two-pin nets. (3) The global router fully utilizes electrically equivalent pins to minimize the routing length of a net. (4) The global router is independent of the layout style since the only inputs to the algorithm are a net list and a channel graph. (5) The global router seeks to minimize the sum of the routing lengths of all of the nets subject to the satisfaction of the capacity constraints of the edges.

9.2 Future Work

The research results presented here on placement and global routing algorithms using simulated annealing have spawned a number of open research areas. This section outlines some of those topics which would lead to improved algorithms.

9.2.1 Simulated annealing

The simulated annealing algorithm is very easily stated and it is quite simple to obtain an implementation for many combinatorial optimization problems. In many cases, it is possible to obtain a good solution by means of the simulated annealing algorithm for a problem in which it is very difficult to develop a direct, or nonprobabilistic, heuristic due to very many degrees of freedom. However, the quality of the solutions obtained by simulated annealing can vary dramatically from one implementation to another. Unfortunately, much experimentation is usually required to develop a high performance simulated annealing algorithm for a new application. This is due to the lack of definitive theoretical results concerning guidelines for optimal properties and control of a simulated annealing algorithm.

Another area for improvement in simulated annealing is knowledge of how to best accommodate multiple terms in a cost function. Simulated annealing provides a high degree of flexibility since there is no operational limit on the number of independent terms in the cost function. Usually an overall cost function is composed of a primary term, which is to be minimized, and then a number of secondary terms, or penalty terms, which should be reduced to zero at the conclusion of the execution of the algorithm. The relative weighting of these terms is known to have a dramatic effect on the quality of the final solution, that is, the final value of the primary term. Too much weight on the primary term is likely to result in a final solution which has some infeasibilities due to a penalty term which failed to be reduced to a sufficiently low value. On the other hand, too little weight on the primary term often yields poor final solution quality. As yet, there are no known definitive guidelines on relative weighting schemes for the various terms.

9.2.2 Row-based cell placement

TimberWolfSC has yielded excellent circuit densities in comparison to other automatic, and even manual methods. The next major issue which needs to be addressed in terms of the quality of the solution yielded by automatic placement and routing is circuit performance. The placement as well as the interconnection patterns play a major role in the performance of a circuit. Incorporating an approximate timing analysis methodology into the simulated annealing framework appears to be a promising approach. This would enable area minimization and performance optimization to proceed simultaneously.

Another area for improvement is execution time. The quality of the results produced by TimberWolfSC are quite satisfactory in comparison to other methods. However, a price is paid in terms of CPU time. A parallel implementation of TimberWolfSC package would provide dramatic execution-time reductions for row-based placement and global routing. There have been some recent attempts at developing parallel implementations of simulated annealing.[1] [2] [3] Since little emphasis was given to CPU time minimization during the development of TimberWolfSC, it is believed that improvements in the single-processor algorithms could also lead to substantial reductions in the execution time.

Another natural step is the extension to other technologies, for example, the sea-of-gates layout style. The placement strategy for the sea-of-gates problem is substantially more complex than for the more pure row-based styles such as gate array and standard cell. The cells need not be confined to one row, nor need they be rectangular in shape. Also, a given cell may be instantiated in any one of a set of possible shapes and sizes. Routing will be accomplished entirely within the rows if possible. If sufficient track space is not available, an entire row, or at least a portion of a row, must go unused to accommodate the required wiring space. A minimum number of rows must go entirely unused. Portions of rows are left unused only where it is necessary to accommodate the interconnect requirements.

I also believe that TimberWolfSC could be extended for use in module generation. The fundamental units to be handled by the placement algorithms are no longer cells, but *transistors*. Recently, an optimizer for generating array-based modules using simulated annealing was shown to outperform currently available tools.[4]

9.2.3 Row-based global routing

The row-based global router should be extended in several ways. First, it currently does not handle electrically equivalent pins on polysilicon in the best possible fashion. For some technologies, due to high resistivity, only one pin of an electrically equivalent pair can be used to complete the routing. This restriction is not currently supported.

The selection of implicit feed-through cross-over points on the rows is not optimized sufficiently. An algorithm should be developed which optimizes the positions of the cross-over points by permuting their locations. The TimberWolfMC global router, since it handles electrically equivalent pins, could also be extended for use in the initial selection of the cross-over points.

---

[1] A. Casotto, F. Romeo, and A. Sangiovanni-Vincentelli, "A Parallel Simulated Annealing Algorithm for the Placement of Macro-Cells," (1987).

[2] S. Kravitz and R. Rutenbar, "Placement by Simulated Annealing on a Multiprocessor," (1987).

[3] P. Banerjee and M. Jones, "A Parallel Simulated Annealing Algorithm for Standard Cell Placement on a Hypercube Computer," (1986).

[4] S. Devadas and R. Newton, "Topological Optimization of Multiple-Level Array Logic," (1987).

The global router should also be augmented for routing track minimization over logically equivalent sets of pins (that is, in addition to minimization over sets of electrically equivalent pins).

9.2.4 Macro/custom cell placement

Circuit performance optimization would be a key area for future work with respect to macro/custom cell placement. Also, development of a parallel simulated annealing algorithm for macro/custom cell placement is a promising future research direction.[1]

9.2.5 Interconnection length estimation

The models and algorithms for estimating the average interconnection length for both random and optimized placement should be extended to include fixed pad and/or cell restrictions. This is not a difficult extension and it would increase the applicability of the model.

9.2.6 Channel definition

The channel definition algorithm has proven to be very robust for use in placement refinement. It should be interfaced to a detailed router. Also, a relatively easy extension to permit routing over the macro/custom cells remains to be done.

9.2.7 Graph-based global routing

As presented in Section 8.3, the total number of routes generated for a net is given by $M_0^{G_n+2}$, where only the $M \geq M_0$ shortest are to be retained and where G_n is the number of pins in the net. Nets with large G_n strongly impact the computation time under the current implementation. It is my opinion that the computation time could be reduced dramatically by slightly changing the manner in which alternative routes are explored. Currently, M_0 routes are generated between every remaining source node and every current target node. During the earlier stages of considering alternatives for a net, when the number of accumulated target nodes is small, it is important to consider many possibilities. However, it is my opinion that as more and more pins become target nodes, then fewer possible connections from the source nodes to the target nodes need to be considered. The quality of the M-shortest routes should not be affected since the number of routes explored for each net would still be dramatically greater than M. Also, the current implementation has the drawback that for very large nets, M_0 must be set to one since even 2^{G_n+2} is too large. Thus very few alternatives are explored for the extremely large nets, thereby explaining the performance problems for these nets. It is my opinion that better performance would be obtained by starting with a value of M_0 larger than one, and then gradually decreasing its value toward 1 as more of the pins are included in the target list.

The combination of the first phase of global router which generates approximately the M-shortest routes for each net and the integer linear programming approach may be particularly effective in selecting the set of shortest routes which satisfy capacity constraints

[1] A. Casotto, F. Romeo, and A. Sangiovanni-Vincentelli, "A Parallel Simulated Annealing Algorithm for the Placement of Macro-Cells," (1986).

on the edges of the channel graph.[1] [2] It would be interesting to compare these results with those produced by the random interchange algorithm which is implemented currently. Also, the random interchange algorithm should be extended to a simulated annealing implementation.

[1] P. Raghavan and C. Thompson, "Probably Good Routing in Graphs: Regular Arrays," (1985).

[2] A. Ng, P. Raghavan, and C. Thompson, "Experimental Results for a Linear Program Global Router," (1987).

Appendix

Island-Style Gate Array Placement

A.1 Introduction

The original program in the TimberWolf package was a generalized *island-style* gate array placement program. This program served to indicate the inherent flexibility of the simulated annealing algorithm. An island-style gate array is such that an array of cells covers the chip with pads on the periphery. Each fundamental unit in the gate array is referred to as a *cell.* Hence, a 50×50 gate array is said to have 2500 cells. A 5×4 island-style gate array is shown in Figure A.1.

Gate array designs often allow additional flexibility and hence greater gate utilization by creating functionally independent units within a cell. For example, some gate arrays utilize functional units which are half-cell sized. TimberWolf handles functional units which are full-cell, half-cell, or quarter-cell sized. The term *module* is used to refer to a functional unit specified in the net list. A module may be the size of: (1) a full cell, (2) a half cell, or (3) a quarter cell. In Figure A.1, cell C_8 contains two half-cell-sized modules. Meanwhile, cell C_{11} contains one half-cell-sized module and two quarter-cell-sized modules. Finally, cell C_{12} contains four quarter-cell-sized modules. A *macro module* consists of a prewired, arbitrarily shaped collection of cells. In Figure A.1, an L-shaped macro module is shown which encompasses cells C_{16}, C_{19}, and C_{20}.

Additional flexibility was introduced by defining *exchange classes* of modules. The modules in an exchange class are only allowed to be interchanged with other modules belonging to the same class. This feature is most often used when a group of modules on the edge of the gate array are to be considered as pad cells. Often the exact location of a given pad is not important, however, it is required that the pad remain on a given edge (or edges) of the chip. Another feature is the ability to restrict the permissible locations for macro modules. For example, it is often the case that gate arrays have wider channels in the center of the array. This is in anticipation of the greatest wiring congestion occurring in this region. Because prewired macro modules usually have fixed cell-to-cell spacing, certain macros may not be placed in the center region (or, on the other hand, the outer regions). TimberWolf allows the designation of cell locations as either suitable or unsuitable for a particular set of macro modules.

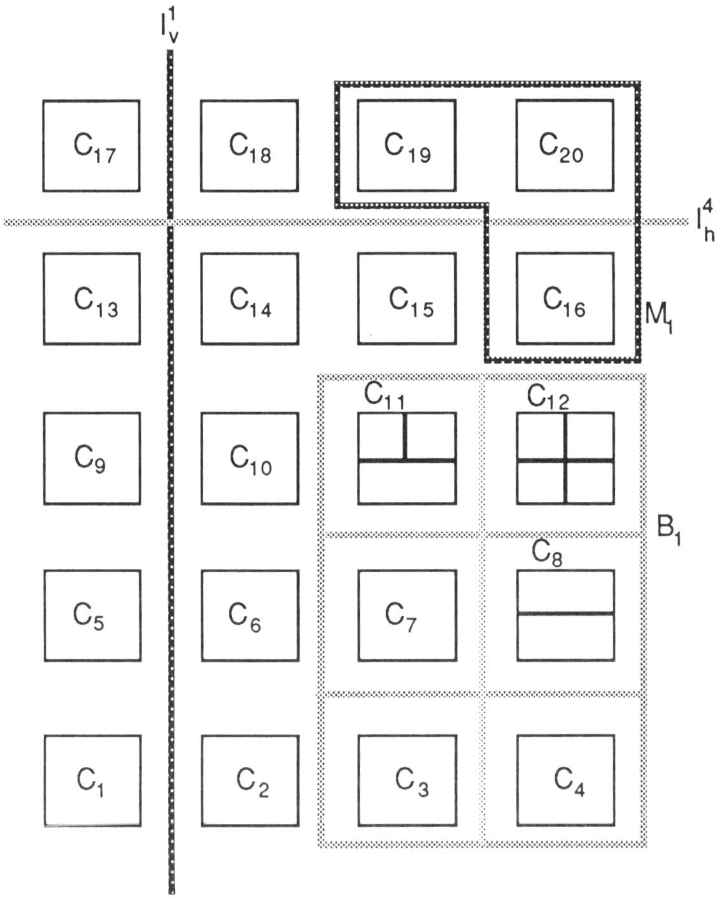

Figure A.1 A 5 × 4 island-style gate array. C_8 contains two half-cell-sized modules. C_{11} contains one half-cell-sized module and two quarter-cell-sized modules. Also, C_{12} contains four quarter-cell-sized modules. An L-shaped macro module is shown which encompasses cells C_{16}, C_{19}, and C_{20}. Vertical channel l_v^1 and horizontal channel l_h^4 are indicated. A total of 17 channel segments are shown in the vicinity of cells C_3, C_4, C_7, C_8, C_{11} and C_{12}.

The TimberWolf gate array placement program can be used with either of two cost functions for the simulated annealing algorithm. The first cost function essentially minimizes the Total Estimated Interconnect Length (TEIL). A second cost function for this program examines the local routing congestion more closely.

The TimberWolf gate array placement program was tested on several standard benchmark placement problems. The placements generated by TimberWolf had final TEIL values which were 6 to 27 percent smaller than the best published results.

This appendix presents the algorithms and the results for the TimberWolf island-style gate array placement program. Section A.2 presents the implementation of the various simulated annealing functions. The results are presented in Section A.3.

A.2 The Implementation of the Simulated Annealing Functions

The simulated annealing algorithmic details for TimberWolf are presented. This section is divided into 5 subsections. Section A2.1 presents the details of the function generate(). In Section A2.2, the cost function is presented. Section A2.3 describes the implementation of the inner loop criterion. Section A2.4 presents the control of the temperature parameter T. Finally, Section A2.5 presents the stopping criterion.

A.2.1 The generation of new states

New state generation is performed by the function generate(). Note that cells may not be fully utilized, or even utilized at all. For example, a cell may only contain one half-cell-sized module, or a cell may contain no modules. The function $C_t(i)$ returns a code which indicates the *type* of a given cell in the array.

$$C_t(i) = \begin{cases} 1 & \text{if 0 or 1 full-cell-sized modules in cell } i \\ 2 & \text{if 1 or 2 half-cell-sized modules in cell } i \\ 3 & \text{if 1 half-cell-sized and 1 or 2 quarter-cell-sized modules} \\ 4 & \text{if only quarter-cell-sized modules in cell } i \end{cases} \qquad \text{(A.1)}$$

A cell j of type $C_t(j) = 1$ has either one full-cell-sized module or one full-cell-sized *empty* module. The function of the empty module is to permit the interchange of a full-cell-sized module into the slot currently occupied by the empty module. A cell j with $C_t(j) = 2$ with only one module is said to contain one module and one empty module. That is, this cell can accommodate another half-cell-sized module. The function of the empty module is to permit the interchange of a half-cell-sized module into the slot currently occupied by the empty module. Similarly, a cell j with $C_t(j) = 3$ with only one module (which must be half-cell sized or else the cell would be characterized by $C_t(j) = 4$) is said to have one module and two empty modules, each of which is quarter-cell sized. A cell j with $C_t(j) = 3$ with two modules also has one empty module which is quarter-cell sized. Finally, a cell of type $C_t(j) = 4$ has a total of 4 modules, from which as many as 3 may be empty modules.

A macro module is specified by an $m \times n$ array of cells, where not every cell in the array need be a part of the configuration of the macro module. Thus, macro modules can have many possible shapes depending on which cells in the configuration of the macro are actively part of the macro. Each macro module has an associated $m \times n$ array of bits, either

0 if the cell is not part of the macro or 1 if the cell is part of the macro. Two macro modules are said to have the same *macro type* if the dimensions of their respective configuration arrays are the same, and further, if their respective associated array of active bits are also the same. The function $M_t()$ returns the type of a macro module. For example, in Figure A.1, an L-shaped macro module M_1 composed of cells C_{16}, C_{19}, and C_{20} is shown. M_1 is a 2 × 2 macro with one inactive cell position, currently occupied by C_{15}. The function $M_C(i)$ returns the macro module to which cell i belongs. If cell i is not part of a macro, the returned value is zero.

The function generate() is shown below:

```
generate()
    do {
        i = R( 1 , N_c ) ;
        j = R( 1 , N_c ) ;
            /* The function R(k,l) returns a pseudo-random integer between k and l, with
            uniform distribution. Cells i and j are randomly selected. Without loss of gen-
            erality, assume C_t(i) ≥ C_t(j). If not, interchange the roles of i and j.
                The variable valid_move is set to FALSE if: (a) there are no modules
            present in either cell, (b) the exchange class of the modules in either cell is such
            that the respective module(s) must remain fixed in the particular cell, or (c) the
            exchange classes of the modules in the two cells differ, thus prohibiting an
            interchange of modules between the two cells. */
    } while( valid_move == FALSE ) ;
    if( M_C(i) > 0 && M_C(j) > 0 && M_C(i) ≠ M_C(j) && M_t(i) == M_t(j) ) {
            /* The cells i and j are each part of different macro modules of the same macro
            type. A new state is generated by interchanging the two macro modules. */
        interchange_macros( M_C(i) , M_C(j) ) ;
    } else if( M_C(i) > 0 && M_C(j) > 0 && M_C(i) == M_C(j) ) {
            /* The two cells belong to the same macro module. The next new state is gen-
            erated by interchanging the two cells (within the macro module), if net list
            specifications permit interchanges within a macro.
        interchange_within_macro( i , j ) ;
    } else if( M_C(i) > 0 || M_C(j) > 0 ) {
            /* One of the selected cells is part of a macro module and the other not. With-
            out loss of generality, assume M_C(i) > 0. A new state is generated by inter-
            changing the macro module M_C(i) with the cells in the vicinity of cell j, where
            cell j forms the upper left corner cell of the m X n array of cells for the macro
            module. If any of the cells in the vicinity of cell j which are to be involved in
            the interchange cannot be moved, or if the macro module is designated as
            unsuitable for placement in that part of the chip occupied by cell j, then a new
            state is not generated. */
        move_macro( M_C(i) , j );
    } else {        /* neither cell belongs to a macro */
        if( C_t(i) == 1 && C_t(j) == 1 ) {
                /* Attempt an interchange of the contents of the two cells. */
            interchange_two_cells( i , j ) ;
        } else if( C_t(i) == 1 || C_t(j) == 1 ) {
```

```
                /* One of the cells has only full-cell-sized modules, and the other con-
                tains multiple modules. Attempt an interchange of the contents of the
                two cells. */
            interchange_two_cells( i , j ) ;
    } else {
            if( C_t(i) == 2 && C_t(j) == 2 ) {
                        /* One module is randomly selected from each of cell i and
                        cell j. The new state is generated by the interchange of the
                        two selected modules. */
                    interchange_two_modules_2_2( i , j ) ;
            } else if( C_t(i) == 3 && C_t(j) == 2 ) {
                        /* One module is randomly selected from cell j. The new
                        state is generated by an interchange of the selected module with
                        the half-cell-sized module of cell i. */
                    interchange_two_modules_3_2( i , j ) ;
            } else if( C_t(i) == 3 && C_t(j) == 3 ) {
                        /* There are two types of interchanges possible: (1) An inter-
                        change of a randomly selected half-cell-sized module from one
                        cell with the 2 quarter-cell-sized modules from the other cell.
                        Note that this changes the type of the former cell to type 4 and
                        the type of the latter cell to type 2. (2) An interchange of a
                        randomly selected quarter-cell-sized module from i with a ran-
                        domly selected quarter-cell-sized module from j. Note that this
                        maintains (at three) the cell types of both i and j. A random
                        selection of (1) or (2) is made. */
                    k = R(1,2) ;
                    interchange_two_modules_3_3( i , j ) ;
            } else if( C_t(i) == 4 && C_t(j) == 4 ) {
                        /* Two new states are generated for this case, where each is
                        generated in the following manner: A quarter-cell-sized module
                        is randomly selected from i and a quarter-cell-sized module is
                        randomly selected from j. A new state is generated by the in-
                        terchange of the two randomly selected modules. */
                    for( l = 1 ; l ≤ 2 ; l++ ) {
                            interchange_two_modules_4_4( i , j ) ;
                    }
            } else if( C_t(i) == 4 && C_t(j) == 2 ) {
                        /* The generation of a new state begins by first randomly se-
                        lecting a half-cell-sized module from j and then by randomly
                        selecting two quarter-cell-sized modules from i. The inter-
                        change of the randomly selected half-cell-sized module with the
                        2 quarter-cell-sized modules constitutes the generation of a new
                        state. Note that if this new state is accepted then the cell types
                        of both i and j change to type 3. */
                    interchange_two_modules_4_2( i , j ) ;
            } else if( C_t(i) == 4 && C_t(j) == 3 ) {
                        /* Two new states are generated for this case, where each is
                        generated in the following manner: A quarter-cell-sized module
                        is randomly selected from i and a quarter-cell-sized module is
                        randomly selected from j. The interchange of the two ran-
                        domly selected modules determines the new state. */
                    for( l = 1 ; l ≤ 2 ; l++ ) {
                            interchange_two_modules_4_3( i , j ) ;
```

```
                }
            }
            interchange_two_cells( i , j ) ;
                /* The next new state is generated by interchanging the contents of the
                two cells. That is, all of the modules in i are interchanged with all of
                the modules in j. */
        }
    }
}
```

A.2.2 The cost function

The estimation of the interconnect length for a single net is determined by computing the half-perimeter of the bounding box for the net. The bounding box is defined by the smallest rectangle which encloses all of the pins comprising the net. For nets comprised of two or three pins, this is the Manhattan interconnect distance. The pins for all of the nets are assumed to lie at the center of their respective cells.

The TimberWolf gate array placement program can be used with either of two cost functions.

A.2.2.1 The first cost function

The first cost function is based on the computation of net-crossing histograms for each horizontal and vertical channel of the placement region. Vertical channel l_v^1 and horizontal channel l_h^4 are indicated in Figure A.1. The histograms are computed by considering the bounding box of each net and adding 1 to the histogram for each channel intersecting the bounding box. The sum of the histogram values for each horizontal and vertical channel is equivalent to summing the half-perimeters of the bounding boxes of each net. Further, a net-crossing threshold value is assigned to each channel. If the number of nets crossing a channel exceeds the specified threshold value, a penalty is assessed proportional to the square of the number of net crossings exceeding the threshold. The threshold mechanism has the effect of evening out the wiring congestion during the earlier stages of the annealing. This has shown to result in a lower value of the total wire length. A partitioning effect may be produced by setting the threshold of a particular channel to zero or a negative value. In this case, nets crossing this channel will be severely penalized. The formulation of the cost function in terms of net-crossing histograms and threshold values was first introduced by Kirkpatrick, Gelatt, and Vecchi.[1]

Let N_V be the number of vertical channels and N_H be the number of horizontal channels. Formally, the value returned by the first cost function is given by:

$$C_1 = \sum_{l_v = 1}^{N_V} \left\{ c(l_v) + p(l_v) \right\} + \sum_{l_h = 1}^{N_H} \left\{ c(l_h) + p(l_h) \right\} \qquad \text{(A.2)}$$

[1] S. Kirkpatrick, C. Gelatt, and M. Vecchi, "Optimization by Simulated Annealing," (1983).

where the function *c()* returns the number of nets crossing the channel specified by its argument, and where the penalty function *p(l)* for channel *l* is given by:

$$p(l) = \begin{cases} (c(l) - t(l))^2 & \text{if } c(l) > t(l) \\ 0 & \text{if } c(l) \le t(l) \end{cases} \tag{A.3}$$

The function *t()* returns the specified threshold limit for the number of nets crossing the channel given by its argument.

A.2.2.2 The second cost function

The second cost function examines the local routing congestion more closely. For this cost function, each *channel segment* is assigned a threshold value. For example, in Figure A.1, 17 channel segments are shown in the vicinity of cells C_3, C_4, C_7, C_8, C_{11} and C_{12}. A channel segment is a portion of a horizontal or vertical channel with a length equal to the center-to-center cell spacing in that region of the array.

Suppose a net interconnects cells C_3 and C_{12}. The bounding box of this net (B_1) is shown in Figure A.1. A total of 17 segments are enclosed by B_1. The congestion per channel segment introduced by this net is approximated as the half perimeter of the bounding box (5) divided by the total number of segments enclosed (17). The factor of 5/17 is the estimated probability of occupancy for the given net in each of the 17 segments. The given net contributes zero to all other segments. The summation of the occupancy probabilities over all nets for a given segment is an estimate of the number of wiring tracks required. The cost function is then the sum of the expected occupancy of each segment plus a penalty assessed for each segment which has occupancy exceeding the corresponding threshold. Specifying a threshold value for each channel segment which reflects the actual fixed channel width increases the likelihood that the final placement will be routable. Furthermore, the total estimated interconnect length will be minimized within the limits of these constraints.

More formally, the second cost function returns the following value:

$$C_2 = \sum_{v_s=1}^{N_s^v} \{O_p(v_s) + p(v_s)\} + \sum_{h_s=1}^{N_s^h} \{O_p(h_s) + p(h_s)\} \tag{A.4}$$

where the function $O_p()$ returns the sum over all nets of the occupation probabilities for the channel segment specified by its argument. N_s^v is the total number of vertical segments and N_s^h is the total number of horizontal segments. The penalty function *p()* is given by

$$p(s) = \begin{cases} (O_p(s) - t(s))^2 & \text{if } O_p(s) > t(s) \\ 0 & \text{if } O_p(s) \le t(s) \end{cases} \tag{A.5}$$

The function *t()* returns the specified threshold limit for the sum over all nets of the occupation probabilities for the channel segment specified by its argument.

More explicitly, with N_n representing the number of nets, the function $O_p()$ is given by

$$O_p(s) = \sum_{n=1}^{N_n} \left\{ o_p(n, s) \right\} \tag{A.6}$$

where

$$o_p(n, s) = \begin{cases} \dfrac{B_p(n)}{N_s(n)} & \text{if } s \in B_s(n) \\ 0 & \text{otherwise} \end{cases} \tag{A.7}$$

The function $B_s()$ returns a list of channel segments contained by the bounding box of the net specified by its argument. The function $B_p()$ returns the half perimeter of the bounding box of the net in terms of a number of channel segments. Finally, the function $N_s()$ returns the total number of channel segments contained by the bounding box of the net.

A.2.3 The inner loop criterion

The inner loop criterion in TimberWolf specifies the number of iterations of the inner loop, that is, the number of executions of the function generate() per value of T, represented by A_c. Thus the number of iterations of the inner loop for a given value of T is given by:

$$A = A_c \cdot N \tag{A.8}$$

where N is the number of cells.

A.2.4 The control of T

This subsection describes the TimberWolf implementation of the update function:

$$T_n = \text{update}(T) = \alpha \cdot T, \qquad 0 < \alpha < 1 \tag{A.9}$$

where α is a constant. The initial value of $T = T_\infty$ is set so that the vast majority of the new states are accepted.

The effect of nonzero values for T_∞ and α can be seen in Table A.1. The circuit used for the test required the placement of 151 (full-cell-sized) modules on an 11×15 array of cells.[1] The value of A_c was set to 30, the value of T_∞ was set to 40.0, and the value of α was set to 0.90.

No. Trials	Annealed Final TEIL	Nonannealed Final TEIL
8	1731-1798	1871-2002
Avg.	1762	1910

Table A.1 The effect of nonzero T_∞ and α.

[1] J. Stevens, "Fast Heuristic Techniques for Placing and Wiring Printed Circuit Boards," (1972).

The term *annealed* in Table A.1 refers to the fact that nonzero values for T_∞ and α were used. Meanwhile, the term *nonannealed* implies that $T = 0$ was maintained throughout the trial. Note that the range of values for the final TEIL in the annealed case was fully less than the range for the nonannealed case. The $T = 0$ case is simply a random interchange algorithm. The same inner loop criterion (specified by the value of A_c) and the same stopping criterion were used for both the simulated annealing algorithm (nonzero value of T_∞) and the random interchange algorithm. The stopping criterion was satisfied if four consecutive iterations ended with the same value of the cost function. For this problem, simulated annealing performed better than the random interchange algorithm for every trial. Note further that the average value for the final TEIL in the annealed case was 8 percent less than that for the nonannealed case.

The effect of α on the final value of the TEIL can be seen in Table A.2, where the circuit of Figure A.2 was used.

α	No. Trials	Average Final TEIL	σ	Worst Solution	No. Optimal
0.95	40	39.60	1.26	44	7
0.99	40	38.75	0.90	41	20

Table A.2 The effect of α.

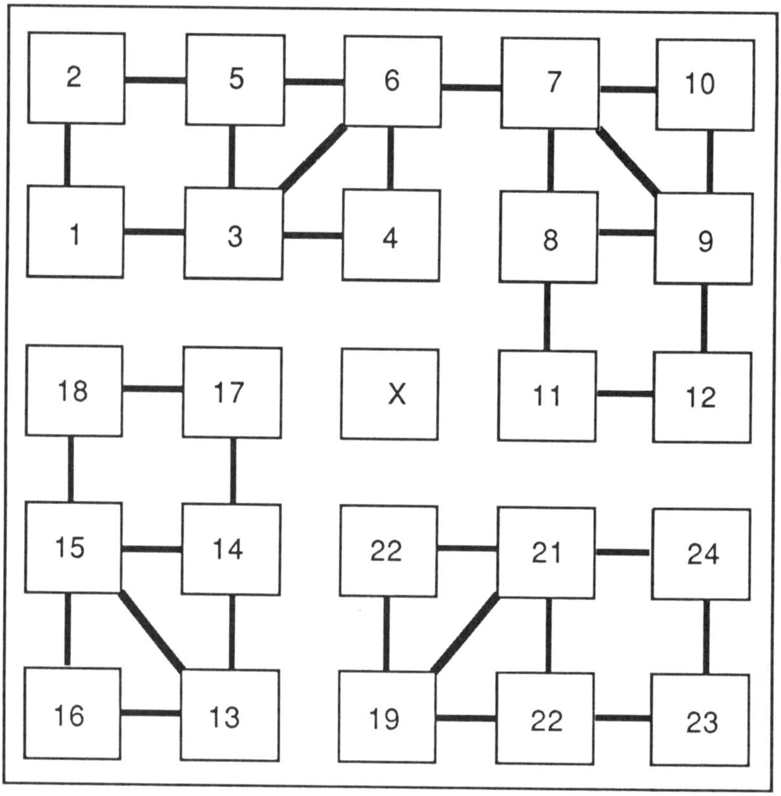

Figure A.2 This circuit contains 24 modules to be placed on a 5×5 grid. The net list is shown by means of the edges interconnecting the squares (which represent the modules). There are a total of 34 nets. There are 16 optimal configurations and the TEIL for an optimal state is 38.

This circuit contains 24 modules to be placed on a 5×5 grid. The net list is shown by means of the edges interconnecting the squares (which represent the modules). There are a total of 34 nets. There are 16 optimal configurations and the TEIL for an optimal state is 38.

The value of A_c was set to 30 and the value of T_∞ was set to 40.0. Note that the average value of the final TEIL is about 3 percent less for the $\alpha = 0.99$ case. Furthermore, the standard deviation was substantially smaller for the $\alpha = 0.99$ case. The number of times the optimal value for the TEIL was achieved was 50 percent of the number of trials for $\alpha = 0.99$, compared to only 18 percent for $\alpha = 0.90$. It is apparent that larger values of α are more likely to result in better final values for the TEIL.

A.2.5 The stopping criterion

The implementation of the *stopping criterion* is very straightforward. The *stopping criterion* is satisfied if the value of the cost function remains unchanged for four consecutive iterations of the inner loop, that is, for four consecutive values of T. Furthermore, to prohibit early, unexpected terminations, TimberWolf does not check if cost has remained unchanged for four consecutive iterations until the condition $T \leq 0.4$ is also satisfied.

A.3 Results

A.3.1 Performance comparison of the two cost functions

The circuit shown in Figure A.2 was used to compare the relative performance of the two cost functions. TimberWolf was tested on this circuit with the first cost function as well as with the second cost function. The performance-comparison data is shown in Table A.3.

Cost Function	α	Average Final TEIL	No. Optimal (percent)
No. 1	0.95	39.60	17.5
No. 1	0.99	38.75	50.0
No. 2	0.97	39.50	40.0

Table A.3 The comparison of the two cost functions.

The first two lines in the table show the data for the first cost function in which the threshold was set to 10.0 for each vertical and horizontal channel. The third line in Table A.3 shows the data for the second cost function in which the threshold was set to 3.0 for each channel segment. In each case, $T_\infty = 40.0$.

The performance of the second cost function is such that in terms of the final TEIL and percentage of optimal final states, it appears to perform nearly as well as the first cost function (taking into account the rate of reduction of T, that is, α). Furthermore, the second cost function kept the local congestion (that is, the expected number of nets passing through each channel segment) at or below 3.0. This second cost function deserves further study. The CPU time necessary to evaluate the second cost function is on the order of 3 to 5 times more than that of the first cost function. However, the additional time may yield substantially improved wireability of the gate array.

A.3.2 Performance comparison on benchmark problems

To test the program and compare it with existing placement techniques, a set of standard benchmarks have been considered. These benchmarks are the ILLIAC IV computer boards reported by Stevens.[1] Note that the printed circuit board problem as stated for

[1] J. Stevens, "Fast Heuristic Techniques for Placing and Wiring Printed Circuit Boards," (1972).

these examples is a particular case of the general gate array placement problem described in the previous subsection.

Three of the ILLIAC IV computer boards were tested. (1) The largest example required the placement of 151 modules on an 11×15 board. TimberWolf reduced the total wire length by 21 percent over Stevens' result and by 17 percent over the result published by Goto and Kuh.[1] (2) The second example required the placement of 108 modules on an 8×15 board. TimberWolf reduced the total wire length by 27 percent over the result published by Goto and Kuh. (3) The third example required the placement of 67 modules on a 5×15 board. TimberWolf reduced the total wire length by 17 percent over Stevens' result and 6 percent over the result published by Goto and Kuh. The value of α remained at a constant value of 0.90 for each of the examples. The results are summarized in Table A.4. CPU times are for a VAX 11/780 running UNIX.

Circuit (No. Modules)	Stevens	Goto and Kuh	TimberWolf	CPU Time in Mins.
151	2181	2098	1731	15
108	untested	1242	909	10
67	700	618	580	5

Table A.4 Comparative results in terms of final TEIL.

[1] S. Goto and E. Kuh, "An Approach to the Two-Dimensional Placement Problem in Circuit Layout," (1978).

Bibliography

Aarts, E. and P. Laarhoven. "Statistical Cooling: A General Approach to Combinatorial Optimization Problems." *Philips Journal of Research* 40 (1985): 193.

Aho, A., J. Hopcroft, and J. Ullman. *The Design and Analysis of Computer Algorithms.* Addison Wesley (1974).

Banerjee, P. and M. Jones. "A Parallel Simulated Annealing Algorithm for Standard Cell Placement on a Hypercube Computer." *Proc. Int. Conf. on Computer-Aided Design* (1986): 34.

Binder, K. *Monte Carlo Methods in Statistical Physics.* New York: Springer-Verlag (1978).

Blanks, J. "Initial Placement of Gate Arrays Using Least-Squares Methods." *Proc. 21st Design Automation Conference* (1984): 670.

Breuer, M. "Min-Cut Placement." *Journal of Design Automation and Fault Tolerant Computing* 1/4 (1977): 343-362.

Burns, J. and R. Newton. "SPARCS: A New Constraint-Based IC Symbolic Layout Spacer." *Proc. Custom Integrated Circuits Conference* (1986): 534.

Casotto, A., F. Romeo, and A. Sangiovanni-Vincentelli. "A Parallel Simulated Annealing Algorithm for the Placement of Macro-Cells." *IEEE Trans. on Computer-Aided Design* 6/5 (1987): 838.

Chen, N. P. "New Algorithms for Steiner Tree on Graphs." *Proc. Int. Symp. on Circuits and Systems* (1983): 1217-1219.

Chen, N. P., C. P. Hsu, and E. S. Kuh. "The Berkeley Building-Block (BBL) Layout System for VLSI Design." in *Proc. Int. Conf. on VLSI.* Amsterdam: North-Holland (1983): 37-44.

Chen, N. P., C. P. Hsu, E. S. Kuh, C. C. Chen, and M. Takahashi. "BBL: A Building-Block Layout System for Custom Chip IC Design." *Proc. Int. Conf. on Computer Aided Design* (1983): 40-41.

Chen, N.P. "Routing System for Building Block Layout." *Ph.D. Dissertation*, UC Berkeley (1983).

Cheng, C. K. and E. S. Kuh. "Module Placement Based on Resistive Network Optimization." *IEEE Trans. on Computer-Aided Design* 3/3 (1984): 218.

Dai, W. M., T. Asano, and E. S. Kuh. "Routing Region Definition and Ordering Scheme for Building-Block Layout." *IEEE Trans. on Computer-Aided Design* 4/3 (1985): 189.

DeCamp, W., G. Sporzynski, and H. Burbank. "Gate Array and Standard Cell Approach." in *Design Methodologies*, ed. by S. Goto. Amsterdam: North-Holland (1986).

Deutsch, J. Private Communication, (1984).

Devadas, S. and R. Newton. "Topological Optimization of Multiple-Level Array Logic." *IEEE Trans. on Computer-Aided Design* 6/6 (1987): 915.

Dijkstra, E. "A Note on Two Problems in Connection with Graphs." *Numer. Math.* 1 (1959): 269-271.

Donath, W. "Placement and Average Interconnect Lengths of Computer Logic." *IEEE Trans. Circuits and Systems* 26 (1979): 272-277.

Donath, W. "Statistical Properties of the Placement of a Graph." *SIAM J. Appl. Math.* 16/2 (1968): 376-387.

Dunlop, A. and B. Kernighan. "A Procedure for Placement of Standard-Cell VLSI Circuits." *IEEE Trans. on Computer-Aided Design* 4/1 (1985): 92-98.

El Gamal, A. and Z. Syed. "A Stochastic Model for Interconnections in Custom Integrated Circuits." *IEEE Trans. on Circuits and Systems* 28 (1981).

Feller, W. *An Introduction to Probability Theory and Applications*. New York: John Wiley (1970).

Fiduccia, C. and R. Mattheyses. "A Linear Time Heuristic for Improving Network Partitions." *Proc. 19th Design Automation Conference* (1982): 175.

Freedman, D. *Markov Chains*. San Francisco: Holden-Day (1971).

Garey, M. and D. Johnson. *Computers and Intractability: A Guide to the Theory of NP-Completeness*. San Francisco: Freeman (1979).

Geman, D. and S. Geman. "Stochastic Relaxation, Gibbs Distributions, and the Bayesian Restoration of Images." *IEEE Trans. Pattern Analysis and Machine Intelligence* 6 (1984): 721-741.

Gidas, B. "Non-Stationary Markov Chains and Convergence of the Annealing Algorithm." *Journal of Stat. Physics* 39 (1985): 73-131.

Gilbert, E. "Random Minimal Trees." *SIAM J. Appl. Math.* 13 (1965): 376-387.

Goto, S. and E. Kuh. "An Approach to the Two-Dimensional Placement Problem in Circuit Layout." *IEEE Trans. on Circuits and Systems* 25/4 (1978): 208.

Goto, S. and T. Matsuda. "Partitioning, Assignment, and Placement." in *Layout Design and Verification*, ed. by T. Ohtsuki. Amsterdam: North-Holland (1986).

Hajek, B. "Cooling Schedules for Optimal Annealing." Manuscript To Appear.

Hakimi, S. "Steiner's Problem in Graphs and its Implications." *Networks* 1 (1971): 113-133.

Hamachi, G. and J. Ousterhout. "A Switchbox Router with Obstacle Avoidance." Proc. 21st Design Automation Conference (1984): 173.

Hanan, M. and J. Kurtzberg. "Placement Techniques." *Design Automation of Digital Systems: Theory and Techniques*. Englewood Cliffs, NJ: Prentice-Hall (1972): 213-282.

Hanan, M., P. Wolff, and B. Agule. "Some Experimental Results on Placement Techniques." *Proc. 13th Design Automation Conference* (1976): 214-224.

Heller, W., W. Mikhail, and W. Donath. "Prediction of Wiring Space Requirements for LSI." *Journal of Design Automation and Fault-Tolerant Computing* (1978): 117.

Hildebrandt, T. "An Annotated Placement Bibliography." *ACM SIGDA Newsletter* (Dec. 1985): 12-21.

Horng, C. and M. Lie. "An Automatic/Interactive Layout Planning System for Arbitrarily-Sized Rectangular Building Blocks." *Proc. 18th Design Automation Conference* (1981): 293.

Hsu, C. P. "A New Two-Dimensional Routing Algorithm." *Proc. 19th Design Automation Conference* (1982): 46.

Huang, M., F. Romeo, and A. Sangiovanni-Vincentelli. "An Efficient General Cooling Schedule for Simulated Annealing." *Proc. Int. Conf. on Computer-Aided Design* (1986): 381.

Iosifescu, M. *Finite Markov Processes and their Applications.* New York: John Wiley (1980).

Isaacson, D. and R. Madsen. *Markov Chains: Theory and Applications.* New York: John Wiley (1976).

Jepsen, D. and C. Gelatt. "Macro Placement by Monte Carlo Annealing." *Proc. Int. Conference on Computer-Aided Design* (1983): 495-498.

Joobbani, R. *An Artificial Intelligence Approach to VLSI Routing.* Boston: Kluwer Academic (1986).

Joobbani, R. and D. Siewiorek. "WEAVER: A Knowledge-Based Routing Expert." *Proc. 22nd Design Automation Conference* (1985): 266.

Kajitani, U. "Order of Channels for Safe Routing and Optimal Compaction of Routing Area." *IEEE Trans. on Computer-Aided Design* 2 (1983): 293-300.

Kani, K., H. Kawanishi, and A. Kishimoto. "ROBIN: A Building Block LSI Routing Program." *Proc. Int. Symp. on Circuits and Systems* (1976): 658-661.

Karlin, S. *A First Course in Stochastic Processes.* New York: Academic Press (1973).

Karp, R., R. Leighton, R. Rivest, C. Thompson, U. Vazirani, and V. Vazirani. "Global Wire Routing in Two Dimensional Arrays." *Annual Symposium on Foundations of Computer Science* (1983): 453-459.

Kernighan, B. and S. Lin. "An Efficient Procedure for Partitioning Graphs." *Bell System Technical Journal* (1970): 291-307.

Kirkpatrick, S., C. Gelatt, and M. Vecchi. "Optimization by Simulated Annealing." *Science* 220/4598 (1983): 671-680.

Krauskopf, J. Private Communication. (1986).

Kravitz, S. and R. Rutenbar. "Placement by Simulated Annealing on a Multiprocessor." *IEEE Trans. on Computer-Aided Design* 6/4 (1987): 534.

Kruskal, J. "On the Shortest Spanning Subtree of a Graph and the Traveling Salesman Problem." *Proc. Amer. Math. Soc.* 7/1 (1956): 48-50.

Kuh, E. S. and M. Marek-Sadowska. "Global Routing." in *Layout Design and Verification*, ed. by T. Ohtsuki. Amsterdam: North-Holland (1986).

Lam, J. and J. M. Delosme. "Logic Minimization Using Simulated Annealing." *Proc. Int. Conf. on Computer-Aided Design* (1986): 378.

Landman, B. and R. Russo. "On a Pin Versus Block Relationship for Partitions of Logic Graphs." *IEEE Trans. on Computers* 20 (1971): 1469-1479.

Lauther, U. "A Min-Cut Placement Algorithm for General Cell Assemblies Based on a Graph Representation." *Proc. 16th Design Automation Conference* (1979): 1-10.

Lawler, E. *Combinatorial Optimization: Networks and Matroids.* New York: Holt, Rinehart and Winston (1976).

Lee, C. Y. "An Algorithm for Path Connections and its Applications." *IRE Trans. on Electronic Computers* 10 (1961): 346-365.

Lie, M. and C. Horng. "A Bus Router for IC Layout." *Proc. 19th Design Automation Conference* (1982): 129.

Luk, W. K. "A Greedy Switch-Box Router." *CMU Technical Report No. 84-148* (1984).

Lundy, M. and A. Mees. "Convergence of the Annealing Algorithm." *Proc. Simulated Annealing Workshop* (1984).

Marek-Sadowska, M. "Two-Dimensional Router for Double Layer Layout." *Proc. 22nd Design Automation Conference* (1985): 117.

Metropolis, N., A. Rosenbluth, M. Rosenbluth, A. Teller, and E. Teller. "Equation of State Calculations by Fast Computing Machines." *Journal Chem. Phys.* 21/6 (1953): 1087.

Mitra, D., R. Romeo, and A. Sangiovanni-Vincentelli. "Convergence and Finite-Time Behavior of Simulated Annealing." *Proc. 24th Conf. on Decision and Control* (1985).

Ng, A., P. Raghavan, and C. Thompson. "Experimental Results for a Linear Program Global Router." *Computers and Artificial Intelligence* (1987).

Ng, A., P. Raghavan, and C. Thompson. "Specification Languages for Describing Rectilinear Steiner Tree Configurations." *Proc. 23rd Design Automation Conference* (1986): 659.

Otten, R. "Automatic Floorplan Design." *Proc. 19th Design Automation Conference* (1982): 261-267.

Otten, R. and L. van Ginneken. "Floorplan Design using Simulated Annealing." *Proc. Int. Conference on Computer-Aided Design* (1984): 96-98.

Ousterhout, J., G. Hamachi, R. Mayo, W. Scott, and G. Taylor. "Magic: A VLSI Layout System." *Proc. 21st Design Automation Conference* (1984): 152.

Prim, R. "Shortest Connecting Networks and Some Generalizations." *Bell System Technical Journal* 36 (1957): 1389-1401.

Raghavan, P. and C. Thompson. "Multiterminal Global Routing: A Deterministic Approximation Scheme." Manuscript To Appear.

Raghavan, P. and C. Thompson. "Provably Good Routing in Graphs: Regular Arrays." *Proc. 17th ACM Symposium on Theory of Computing* (1985).

Reed, J., A. Sangiovanni-Vincentelli, and A. Santamauro. "A New Symbolic Channel Router: YACR2." *IEEE Trans. on Computer-Aided Design* (1985): 208.

Reif, F. *Statistical and Thermal Physics.* New York: McGraw-Hill (1965).

Rivest, R. "The PI (Placement and Interconnect) System." *Proc. 19th Design Automation Conference* (1982).

Roberts, M. Private Communication. (1986).

Romeo, F. and A. Sangiovanni-Vincentelli. "Probabilistic Hill Climbing Algorithms: Properties and Applications." *Chapel Hill Conference on Very Large Scale Integration.* Computer Science Press (1985): 393-418.

Sangiovanni-Vincentelli, A. Private Communication (1986).

Schwartz, J. "Fast Probabilistic Algorithms for Verification of Polynomial Identities." *Journal of ACM* 27/4 (1980).

Schweikert, D. and B. Kernighan. "A Proper Model for the Partitioning of Electrical Circuits." *Proc. 9th Design Automation Workshop* (1972): 57-62.

Sechen, C. and A. Sangiovanni-Vincentelli. "The TimberWolf Placement and Routing Package." *Proc. Custom Integrated Circuits Conference* (1984).

Sechen, C. and A. Sangiovanni-Vincentelli. "The TimberWolf Placement and Routing Package." *IEEE J. of Solid-State Circuits* 20/2 (1985): 510.

Sechen, C. and A. Sangiovanni-Vincentelli. "TimberWolf3.2: A New Standard Cell Placement and Global Routing Package." *Proc. 23rd Design Automation Conference* (1986): 432-439.

Sechen, C. and K. W. Lee. "An Improved Simulated Annealing Algorithm for Row-Based Placement." *Proc. Int. Conf. on Computer-Aided Design* (1987).

Sechen, C., D. Braun, and A. Sangiovanni-Vincentelli. "ThunderBird: A Complete Standard Cell Layout Package." *IEEE J. of Solid-State Circuits* 23/2 (1988).

Selby, S., Editor. *CRC Standard Mathematical Tables*. Cleveland: CRC Press (1973).

Senata, E. *Non-Negative Matrices and Markov Chains*. New York: Springer-Verlag (1980).

Shin, H. and A. Sangiovanni-Vincentelli. "A Detailed Router Based on Incremental Routing Modifications: MIGHTY." *IEEE Trans. on Computer-Aided Design* 6/6 (1987): 942.

Srinath, G. "NLAGR." Unpublished Manuscript. UC Berkeley.

Steinberg, L. "The Backboard Wiring Problem: A Placement Algorithm." *SIAM Rev.* 3/1 (1961) 37-50.

Stevens, J. "Fast Heuristic Techniques for Placing and Wiring Printed Circuit Boards," *Ph.D. Dissertation*, U. Illinois (1972).

Sutherland, I. and D. Oestreicher. "How Big Should a Printed Circuit Board Be?." *IEEE Trans. on Computers* 22 (1973): 537-542.

Vecchi, M. and S. Kirkpatrick. "Global Wiring by Simulated Annealing." *IEEE Trans. on Computer-Aided Design* 2/4 (1983): 215-222.

White, S. "Concepts of Scale in Simulated Annealing." *Proc. Int. Conference on Computer Design* (1984): 646.

Wong, D. and C. Liu. "A New Algorithm for Floorplan Design." *Proc. 23rd Design Automation Conference* (1986).

Index